PLATYPUS

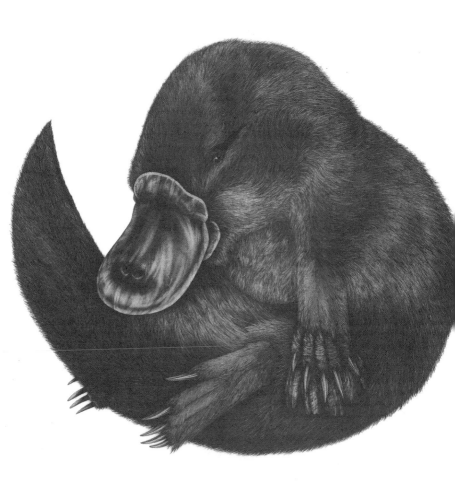

PLATYPUS

THE EXTRAORDINARY STORY OF HOW
A CURIOUS CREATURE BAFFLED THE WORLD

Ann Moyal

Smithsonian Institution Press
Washington, D.C.

First published in 2001

Copyright © Ann Moyal 2001

Published in 2001 in the United States of America
by the Smithsonian Institution Press
in association with Allen & Unwin
83 Alexander Street
Crows Nest, Sydney NSW 2065
Australia

Library of Congress Cataloging-in-Publication Data
Moyal, Ann Mozley.
    Platypus : the extraordinary story of how a curious creature
    baffled the world / Ann Moyal.
        p.   cm.
    Includes bibliographical references (p.).
    ISBN 1-56098-977-7 (alk. paper)
        1. Platypus. I. Title.
    QL737.M72 M68   2001
    599.2'9—dc21   2001020892

Set in 11/14.5 pt Garamond 3 by Bookhouse, Sydney

Line drawings used alongside chapter titles by Nancy Sawer

Endpapers: *Mappa Monde ou description du Globe terrestre assujetie
aux observations astronomiques par le Sr Janvier*, 1804, Nan Kivell
collection, National Library of Australia

Printed in China by Everbest Printing Co. Ltd,
not at government expense.
08 07 06 05 04 03 02 01   5 4 3 2 1

To Jamie—
and other paradoxes

And the world's great story is left untold,
And the message is left unsaid.

A. B. 'Banjo' Paterson,
The Animals Noah Forgot

Platypus.
The first *Ornithorhynchus*
Confused early thinkers.
They said, 'Oh good lord,
It's an obvious fraud!
Somebody has stuck
The front end of a duck
(With the skill of a weaver)
To part of a beaver.
It's no less a fake
Than the mermaids they make
From a fish and an ape—
A ridiculous jape!'

We now know it's real
Though I can't help but feel
That from tail tip to muzzle,
It still is a puzzle.

R. Strahan,
The Incomplete Book of Australian Mammals

CONTENTS

ACKNOWLEDGMENTS

A number of people have given me most valuable assistance in the writing of this book. The Petherick Room of the National Library of Australia has been my hearth, and there I have enjoyed and benefited from the help of a welcoming staff.

The following colleagues and friends have given me timely encouragement, information and creative assistance: Bill Gammage and Graeme Byrne with Aboriginal legends, L. Fisk of the Healesville Sanctuary, Peter Glow, Jamie Mackie, the late Elizabeth Newland, Uwe Proske, Nancy Sawer, Chris Tidemann and Geoff Williams of the Australian Platypus Conservancy.

I am especially grateful to my editors, Colette Vella, Karen Ward and Simone Ford, and to my publisher, Ian Bowring, for their creative and skilful interest and assistance.

I acknowledge with thanks Ronald Strahan's permission to quote from his poem on the *Ornithorhynchus* published in his *The Incomplete Book of Australian Mammals*, Kangaroo Press, Sydney, 1987; the Australian Stockman's Hall of Fame, Longreach, Queensland for the use of Harry Burrell's poem, 'The Mud-Sucking Platypus'; and the John Curtin Prime

Ministerial Library, Curtin University of Technology, Perth, for H. V. Evatt's cablegram to Canberra of May 1943.

Acknowledgment of permission to use copyright pictorial material in the possession of libraries, publishers or individuals is made in 'Illustrations' near the end of the book.

Ann Moyal
Canberra, 2000

INTRODUCTION

F ew people have seen a platypus in the wild. I had not. Writing about its history, I knew I must see the legendary animal in its natural environment. One winter afternoon, at the invitation of the Director of the Australian Platypus Conservancy in Victoria, I set off on the long drive to Healesville and Whittlesea. My journey took me along dappled tree-fringed country roads as labyrinthine, it seemed, as the platypus burrow itself, until afternoon light deepened into early twilight and I came to the appointed reservoir, grey-blue below the mountains.

And there, on the water's rim where roots thickened and overarching gum trees dipped, a small brown head appeared gliding, the body, as one nineteenth-century observer put it, 'flat as a plank' in the water, its presence confirmed by the circle of light concentric ripples in the stream.

I had seen a platypus! Not, as one sometimes sees them, diving fitfully in an aquatic tank at a zoo. But in its peaceful native habitat, cruising quietly in pursuit of its evening meal. After so long and tortuous a drive, magic!

The platypus has become an Australian icon—its small furry frame pressed into a thousand commercial replicas and cast on the back of the twenty-cent coin; its sleek, unmistakable form serving as company logos and a cheerful Australian Olympic symbol; and its name recalled by nostalgic expatriates around the world.

Some of us know that it is a small aquatic mammal which, unlike any other living mammal excepting its fellow monotreme, the echidna, lays eggs.

In April 1999, the story of the birth in captivity of two platypuses at Healesville Sanctuary outside Melbourne caught the attention of the media and the world. Fifty-five years after the first captive birth of this enigmatic animal, a second success had been achieved. One very small male platypus had emerged from the artificial burrow, to be followed, amid surprise and jubilation one week later, by a second healthy platypus, another boy.

In a few short weeks, the event reignited the intense interest that the platypus had evoked following its first appearance in Britain, as a dried specimen, at the end of the eighteenth century. When this first specimen reached England in 1799, with its webbed feet, fur, and a bill like a duck seemingly 'stitched' onto the body, the platypus was regarded as a hoax, a high frolic practised on the scientific community by some colonial prankster. After examining it carefully, dubious naturalists declared it to be real, an animal that lived and breathed. But the platypus continued to confound international biologists for nearly 90 years.

Above all, the nature of its reproductive process remained a mystery. This mystery was crucial. For the late eighteenth and nineteenth centuries saw zoologists, botanists, systematists and theoreticians in scientific centres in Britain and Europe keenly concerned with the ongoing task of refining the classificatory systems of the animal and plant kingdoms. It was a task and a challenge that multiplied as new genera and species, and at times new orders and classes, were brought to light by the exploration of the New World.

Taxonomy and systematics—the identification and classification of plants and animals—had their roots in classical times. A key handful of seventeenth- and eighteenth-century scholars from Britain and the Continent had made distinctive contributions. But it was the nineteenth century that was destined to become the great century of classification and the decoding of the complex, and diversifying, book of Nature.

In this task, the continent of Australia, with its unique fauna and, in a hemisphere far distant from the centres of Western research, had a vital part to play and became a major outpost of information and an important contributor to the theory of evolution.

In the twentieth century, research on the platypus would shift to Australia itself where, across another hundred years or more, the animal was to continue to dazzle and surprise biologists. It became, as one distinguished researcher inscribed it, 'the animal of all time'.

This is its story.

'THIS HIGHLY INTERESTING NOVELTY'

No country ever produced a more extraordinary assemblage of indigenous productions.

P. P. King, *Narrative of a Survey of Intertropical Waters and Western Coasts of Australia Performed Between the Years 1818 and 1822*

The river is very still on the curve where the eucalyptus dips towards the water. The light shades towards late afternoon and twilight will soon darken the outline of the wooded bank and the flat landscape stretching towards the horizon. Bubbles break the surface of the water. A small brown head, its sleek cap visible for no more than two centimetres, glides silently in the river's flow.

Silent, intent on its evening search for food, this is the platypus—the timid, secretive, and highly improbable creature that was to baffle the scientific world for over 90 years, cast

Britain and Europe's leading zoologists into a whirl of fierce dispute, and become a key, if unexpected, player in the great theoretical and philosophical debates that were to transform our understanding of the animal kingdom and usher in a new century of biological science.

Neither fish nor fowl, bird or reptile, the platypus, it seemed, came out of an Antipodean ark, with its webbed feet, duck-like bill, dense glossy fur, high-placed, bright beady eyes, and a small body shaped for swift propulsion through river and lagoon—a living fossil turned up in the distant southern reaches of the globe.

Ever since the expeditions of exploration launched in the sixteenth century, adventurous men had brought back from New World lands exotic creatures that startled the gaze and challenged the understanding of Europeans. The marsupial opossum as early as 1500, and later the giraffe, the hippopotamus, the armadillo and the sloth proved eccentric wonders, defying the naturalists who believed that the Garden of Eden they knew was already full, and dazzling the public through pictorial presentation or through stuffed or living specimens in museums or zoos.

It was, however, the *Endeavour* voyage of Captain Cook, despatched by the British Admiralty in 1768 to observe the transit of Venus at Tahiti and to locate the rumoured Great South Land that, with the multitude of specimens assembled by Joseph Banks, took scientific collection of fauna and flora to a new level. Previously unknown worlds of zoological and botanical species were now thrown open to European view, beginning a transformation in scientific understanding of the natural world.

Cruising along the eastern coast of Australia in 1770 and located for six weeks on north Queensland's Endeavour River, Banks and his Swedish colleague, Carl Solander, had good opportunity to observe the odd flora and fauna that were to brand Australia as an upside-down world of reverses and differences where all seemed 'queer and opposite'. It was they who introduced its rarities to the scientific world. Marooned on the river for six weeks while the damaged *Endeavour* was careened for safe passage through the Great Barrier Reef, the youthful Banks observed a bounding animal that outstripped his fleet greyhounds—a creature, he jotted in his journal, that 'instead of going upon all fours...went only upon two legs, making vast bounds'. Having enjoyed eating the flesh of this remarkable creature later, he and his captain took the skin and skeleton of the kangaroo back to England.

Banks' success as a travelling naturalist and the great assemblage of Australasian and Pacific natural history which the expedition took back to England propelled him to rapid fame and leadership in British science. Appointed director of the King's Royal Botanic Gardens at Kew, he was subsequently elected president of the prestigious Royal Society of London in 1778 at the age of 35, created a baronet a few years later, and, for 40 years until his death, he set his entrepreneurial seal and patronage upon Imperial science. In addition, Banks' recollections of the vegetation near Sydney helped inspire his government's choice of Botany Bay for the establishment of a British penal settlement in 1788. Science followed the flag.

Resultingly, there was a lively interest in any such Antipodean novelties that government appointees and naval men might despatch to England. The Colony's first governor,

Arthur Phillip, set the official ball rolling in 1791 by sending a live kangaroo for King George III. But a people's kangaroo had preceded it.

'The Wonderful Kangaroo from Botany Bay (*the only one ever brought alive to Europe*)', a London poster declared excitedly in 1790, would be 'on exhibition at the Lyceum in the Strand from 8 o'clock in the Morning, till 8 in the Evening'. '*Different from all* QUADRUPEDS', it exclaimed, 'let it suffice to observe that the Ingenious are delighted, and the Connoisseur impressed with Wonder and Astonishment, at the unparalleled animal from the Southern Hemisphere'. Queues crowded to pay and gape for a costly one shilling.

The kangaroo, 'five feet high', 'beautiful' and 'amazing', enthralled Londoners, and the great natural history populariser Thomas Bewick included the new animal in his *General History of Quadrupeds* (1792). But stranger findings would before long appear from New South Wales.

During 1798, Phillip's successor, Governor John Hunter, a keen student of natural history, had watched an Aborigine spearing 'a Small Amphibious Animal of the mole kind' in a lake near the Hawkesbury River close to Sydney. Hunter sent the animal's skin, preserved in a keg of spirits with a Womback [wombat] to the Literary and Philosophical Society of Newcastle-upon-Tyne, which had recently honoured him with corresponding membership. The small creature (Hunter later told the anatomist Everard Home) had fought for its life with such force that it caught its assailant with its spur. Its entry into the scientific community was just as dramatic.

The cask containing the two specimens, so one story ran, reached Newcastle late in 1799 and was transported from the

4

quayside to the Society's rooms by a woman servant. She carried it on her head and, by mischance, the bottom of the cask gave way, dousing her with the pungent spirits. But her dismay and horror was reportedly the greater when looking down she saw not only the small chunky wombat, but the remains of 'a strange creature, half bird, half beast, lying at her feet'.

The platypus had arrived in Britain.

Thomas Bewick, himself a citizen of Newcastle-upon-Tyne, aided by Hunter's notes and drawings introduced the platypus to the general public. The creature, he wrote in the fourth edition of his popular *General History of Quadrupeds* (1800) 'seems to be an animal *sui generis*; it appears to possess a three fold nature, that of a fish, a bird and a quadruped,

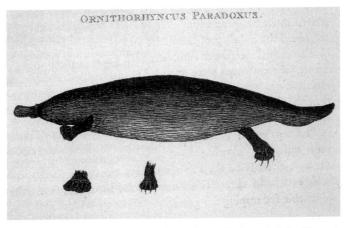

ORNITHORHYNCUS PARADOXUS.

'A Small Amphibious Animal of the Mole Kind.' Captain John Hunter's drawing and description of the then unknown platypus, a dried specimen of which was despatched late in 1798 to Sir Joseph Banks.

5

and is related to nothing that we have hitherto seen'. About the size of 'a small cat', with a bill 'very similar to that of a duck', it had four short legs, 'the fore legs…shorter than those of the hind and their webs spread considerably beyond the claws'. Bewick resisted, as he put it, any attempt 'to arrange it in any of the useful modes of classification' and his benign illustration of the odd little creature reflected none of the struggle that had occasioned its transfer across the world.

Further specimens of the 'amphibious mole' soon began arriving to whet the appetite of British and European naturalists. The eminent naturalist Dr George Shaw, securing a dried specimen during 1799 from an unnamed source in New South Wales, described the physical characteristics of the animal—the body resembling an otter in miniature; 'the head flattish and rather small'; the mouth or snout exactly resembling some broad-billed species of duck; the tail 'flat, furry like the body'; the legs very short and on the forefeet five claws, straight, strong and sharp pointed; on the hind feet six claws, longer and more inclining to a curved form; the length of the whole animal thirteen inches. He published an illustration of the animal by his associate Frederick Nodder, together with drawings of its bill and fore and hind feet, in his *Naturalist's Miscellany* that year, the first published depiction of the platypus. Calling on his knowledge of Greek, Shaw bestowed on the animal the name *Platypus anatinus*, from the Greek *platypous*, meaning flat-footed, and the Latin *anatinus*, meaning duck-like.

But Shaw, an experienced naturalist, a Fellow of the Royal Society and Assistant Keeper of Natural History at the British Museum, clearly regarded the peculiar specimen with suspicion.

Could it, he wondered, be a hoax? Was he being deceived? Chinese and Japanese taxidermists were notorious for the skill with which they constructed nonexistent animals, stitching the head and trunk of monkeys to the hind parts of fish and selling them to credulous sailors. Was this a colonial prank? 'Of all the Mammalia yet known', Shaw wrote in his account, 'it seems the most extraordinary in its conformation; exhibiting the perfect resemblance of the beak of a Duck engrafted on the head of a quadruped'. 'I almost doubted the testimony of my own eyes with respect to the structure of this animal's beak', he mused, and it was only with 'the most minute and rigid examination' that he could persuade himself of its being the real beak or snout of a quadruped. But, he confessed, he could perceive no appearance of any deceptive preparation. 'This paradoxical quadruped', the renowned naturalist concluded wisely, 'must be left to future investigation'.

Yet meticulous and clearly rattled, Shaw returned to the

Was this a colonial prank? Dr George Shaw's platypus illustrated by Frederick Nodder in Naturalists' Miscellany, *1799.*

7

matter in 1800. By then Joseph Banks had received from Governor Hunter two specimens, preserved in spirits, and the reality of Australia's freak of nature was confirmed.

With new confidence, Shaw re-ran his original description of the platypus in his *General Zoology* and declared that it constituted 'a new and singular genus'. Twelve thousand miles from the platypus's habitat, he looked forward to more information from observations in the Colony on the natural history of an animal so astonishingly different from other quadrupeds. This new addition, he added, indeed confirmed strikingly the observation of the great French naturalist Buffon that 'whatever was possible for Nature to produce has actually been produced'.

Banks, scientific patron and promoter, was now firmly in on the act. Sitting at the head of British science, he had received two more specimens from Hunter, both of a platypus skin together with beak and bones. Of these he sent one to Professor Johann Blumenbach, one of Europe's leading comparative anatomists. Blumenbach, based at the University of Göttingen, a foreign Fellow of the Royal Society and a long-time correspondent of Banks, had already got wind of the 'highly interesting novelty' brought over from Botany Bay and begged for a specimen to depict in his German natural history miscellany, *Abbildungen*. Rewarded, he communicated with Banks in some excitement in March 1800, having given the new creature the scientific name *Ornithorhynchus paradoxus*.

'It needs not to tell you', he wrote, 'how exceedingly I must be surprised by the view of so strange a creature as the Ornithorhynchus paradoxus—for being under the necessity of christening it, I thought this a convenient name, taking the generical one from its most remarkable character, the Birdlike Beak & the nomen triviale from its quite paradox shape'.

Blumenbach evidently lacked knowledge of Shaw's description. The winds of international science blew randomly and, for a period, the little animal's two titles stood side by side. But it was found that Shaw's generic name *Platypus* had been used in 1793 for a genus within the order Coleoptera (beetles), and hence could not stand. However, with its apt classical derivation of 'flat-footed', the 'platypus' and its plural 'platypuses' remain the animal's popular moniker. Blumenbach's generic designation passed into science, and in an appropriate amalgamation, Shaw's species name, *anatinus* was added to establish *Ornithorhynchus anatinus* on the zoological register.

In every way a paradox, the Australian arrival raised a host of questions. Was it, as its brown pelt suggested, a mammal? (for possessing a covering of fur or hair was a defining characteristic of the mammalian class). Or was it a reptile (among which amphibian animals were then grouped)? Alternatively, its duck-like bill indicated an affinity with warm-blooded birds. Blumenbach, puzzling, also remembered earlier writings about the griffin, a mythological beast with the body of a lion and the wings and head of an eagle. Plainly no myth, the animal now confronting naturalists combined distinctive parts of mammal, reptile and bird.

Clearly, Blumenbach's *Ornithorhynchus* fitted none of the commonly recognised divisions demarking the major classes of vertebrates—the mammals, fish, birds and reptiles. But, as he outlined in a series of papers published in 1800, the German anatomist considered that the animal conformed to the normal mammalian plan of organisation, and while worried by its beak, he assumed that the female managed to suckle its young in some fashion with its duck-bill.

Blumenbach and Dr Shaw enjoyed a particular advantage. Both had some knowledge of another of Australia's extraordinary fauna—the 'ant-eating porcupine' which had arrived in England from New South Wales some seven years before the platypus. Shaw had described the animal, with illustrations by Nodder, in his *Naturalist's Miscellany* in July 1792, and five years later the great French naturalist Georges Cuvier had placed the porcupine anteater within the mammalian order 'Edentates' (quadrupeds without incisive teeth) where it joined the anteaters, sloths and armadillos, and named it *Echidna hystrix*.

Captain William Bligh of *Bounty* fame had also early recorded the echidna's presence when, on his second tour of duty in Australian waters, he put in at Bruny Island off southeast Tasmania in February 1792, and noted in his logbook that one of his officers had killed an animal 'of very odd form'. 'It had no mouth like any other animal, but a kind of duck bill which opens at the extremity where it will not admit the size of a small pistol. Quills or prickles are all over its back.' Bligh sent a drawing of the eccentric creature to his mentor Banks who doubtless passed it on to his German correspondent.

Banks, meanwhile, had directed the second of Hunter's dried platypus specimens to the distinguished British anatomist Everard Home at the Royal College of Surgeons, London.

Trained at the College under its legendary leader John Hunter, Home was also a Fellow of the Royal Society and one of that breed of naturalists destined to establish a strong discipline of comparative anatomy in Britain. The study of the platypus would occupy him for many years. In his first paper on the animal in 1800, Home focused on the platypus beak and from the dry specimen at his disposal, he made it clear that despite appearances to the contrary, resemblance between it and that of a duck was essentially superficial.

Specimens of the elusive 'duck-bill mole', as the colonists continued to call it, now began to arrive from New South Wales with their organs preserved in spirits, giving eager anatomists access to the internal parts. Hunter's successor, Governor King, supplied Banks with a male and female, 'no pains or spirits', he wrote importantly, being spared to preserve them. From his careful dissections of these prizes, Home was able to provide the first full description of the animal's anatomy. At the same time, he enjoyed the advantage of a first-hand account of the platypus from Governor Hunter on his return from Australia in May 1801.

Hunter, a keen on-the-spot observer of Australian wildlife, had developed his own original ideas on the creation of the curious beasts that seemed so at variance with the northern hemisphere's natural world. In a publication as early as 1793, the handsome Hunter suggested that 'a promiscuous intercourse between the different sexes of all these different animals' might account for their unlikely forms. It was a concept of random promiscuity unlikely to appeal to the knowledgeable Home, but it was shortly taken over by Charles Darwin's lively grandfather, Dr Erasmus Darwin, and published to underpin

Everard Home. His 1802 paper on the anatomy of the platypus remained the foundation scientific source of knowledge about the animal for 25 years.

his theory of evolutionary development from a single form in his *Zoonomia* (1796).

For his part, Home concentrated his sound anatomist's mind on the platypus's beak and its internal organs. In a series of papers written from 1800–02, he found that the beak was not a part of the mouth but, rather, an exploratory organ for touching and tasting that served underwater in place of sight and smell. Its soft fleshy edges functioned as sense organs which enabled the animal to feel in the mud for the water insects and small crustaceans that comprised its food.

In respect of its process of generation, Home noted the remarkable character of the reproductive organs of both female and male platypus and, finding that they were not comparable with those of mammals, examined the corresponding parts of birds and reptiles and found the closest resemblance to the organs of ovoviviparous lizards who produce their young by hatching them from eggs formed inside the mother's body.

He went further. He determined that, again unlike other quadrupeds, the female platypus possessed oviducts that opened with her excretory tract into one common chamber, the cloaca, rather than forming a uterus. The male also expelled its urine through the cloaca from an opening at the base of the penis, reserving the penis for the passage of semen. 'These characteristics', Home summed up, 'distinguish the Ornithor-hynchus, in a very remarkable manner, from all other quadrupeds, giving *this new tribe* a resemblance in some respects to birds, in others, to the Amphibia; so that it may be con-sidered as an intermediate link between the classes of Mammalia'. There was every reason to believe, he concluded, 'that this animal is ovi-viviparous [sic] in its mode of generation'.

But it was no straight run. Blumenbach, Shaw and Cuvier all judged the platypus to be a mammal. Yet the apparent absence of nipples or mammary glands proved a stumbling block for Home. Transitional between bird, reptile and mammal, in its first dissection the paradoxical platypus appeared to inhabit a zoological no-man's land.

The stage was set for conflict. Five years after its discovery

13

in Australia three central questions about the platypus confronted the international fraternity of naturalists and anatomists. First, what light did the strange mix of the platypus's anatomical features shed on the classificatory scheme which naturalists had been designing to embrace the known fauna of the northern world?

Second, how did this curious animal from the Antipodes produce its young? If it was not 'viviparous', producing its young like other mammals, was it in truth 'ovoviviparous' like some lizards with eggs formed and hatched within the female's body? Or was it, perhaps, 'oviparous', hatching its young from eggs laid outside its body, like a bird?

And third, most importantly, what relevance did this anomalous animal, drawn from its remote habitat—this misalliance of God's creatures—have for the old ideas of a perfectly ordered world of creation and the emerging medley of notions about evolution that were beginning to stir in zoological circles as the nineteenth century dawned?

As George Shaw had thoughtfully concluded when several of Australia's indigenous animals landed on his specimen table, the southern continent 'seemed to have been arranged according to a plan different from those that shaped the rest of the globe'.

One man who had instantly seized on the importance of the platypus was the young voyager-turned-scientific patron and diplomat, Sir Joseph Banks. From his influential position, he corresponded with colonial governors, issued instructions to His Majesty's ships' captains for collecting and preserving specimens and spurred major maritime and botanical ventures which opened up a widening knowledge of the scientific riches of the Imperial regions of the world.

During 1800 he had from his own private resources sent George Caley, a botanical collector trained at Kew, to Sydney to begin the systematic collection of Australian plants. Amid growing interest, Banks was soon urging the diligent botanist to extend his skills and observations to the platypus. 'What I want to know', he wrote Caley in August 1802, 'is the mode of generation of the Duck Bill and the Porcupine anteater, also specimens of both animals preserved in spirit which should be once renewed after having stood for a month or two on the animal'. Fired by Home's revelations, he repeated the request urgently the following year. 'Our greatest want here', he pressed early in April, 'is to be acquainted with the manner in which the Duck Bill Animal and the Porcupine Ant Eater which I think is of the same genus, breed, their internal structure is so very similar to that of Birds that I do not think it impossible that they should lay Eggs or at least as Snakes and some Fish do Hatch Eggs in their Bellies'.

Caley faced his new responsibility with customary application, moving out from his depot at Parramatta some 30 kilometres from Sydney in a widening arc. He was the earliest of the Australian naturalists and collectors to cultivate the Aborigines, talking with them, heeding their words politely, and watching them hunt as they burned out the kangaroos from the bush and caught them with their spears.

Caley now made 'strict enquiry' among them and reported that, although he asked several, he could not receive any satisfactory answer. At length he learned from one man 'what

I think may be relied', he wrote Banks in April 1803, who told him that 'they went a long way underground and layed eggs'. 'The natives', he recorded, 'call them by two distinct names, viz. *Bat* and *Malangsing*'. Caley himself admitted that the porcupine anteater was entirely unknown to him, and, from what he learned from the natives, 'is scarce'.

His enquiries clearly stimulated the trade; sellers charged what the market would bear. Caley subsequently bought his porcupine anteater from the Aborigines for five gallons of rum!

Banks was also activating other searchers. Robert Brown, whom he had personally chosen to serve as Matthew Flinders' botanist on his expedition of Australian circumnavigation on HMS *Investigator* in 1801–04 and to record the flora of the new land, was pressed for help. 'I have not yet had an opportunity of dissecting the female ornithorhynchus', Brown responded from New South Wales in August 1803, although a fellow officer had 'found the urethra of the male to terminate in the anus', a fact with which Brown thought Banks was 'probably acquainted, and which, indeed, might almost be inferred from the double-headed penis'. A month later, confronted with a female platypus specimen that was 'unimpregnated and not quite fresh', Brown told his mentor that he had nothing to add 'to your letter respecting the organs of generation except that I found the fallopian tubes completely impervious at their junction with the vagina'. The botanist's data on male and female platypuses confirmed the dissections of Home.

The *Investigator* expedition also captured a special record. The ship's botanical artist, Austrian-born Ferdinand Bauer— described as the 'Leonardo of natural history painting'— carefully made a life-size pencil drawing of two platypuses

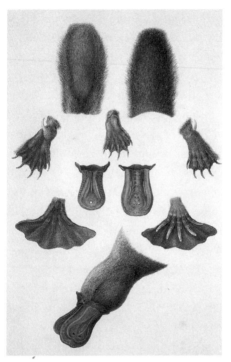

Drawing by Ferdinand Bauer of parts of the anatomy of the platypus, 1802. With his keen scientific eye and delicate brush, Bauer represented the under and upper tail, the claws, snout, webbed feet and head of the extraordinary animal.

observed in New South Wales, sketching them in full and noting their colours against a coded chart of complex hues that he carried with him on the journey, so that he could complete the illustrations on his return to England. His depiction of the animals, with their stout tails and strong beaks, was the first drawing of the live animal since Hunter's rudimentary sketch—and it lay hidden, unpublished by the artist, in the British Museum of Natural History until it was unveiled to public view among Bauer's remarkable floral and natural history drawings in the second-last decade of the twentieth century.

Meanwhile in Australia, in the eastern rivers and creeks of the huge continent stretching from Queensland down the coast of New South Wales and Victoria and across Bass Strait into Tasmania, the platypus itself, untroubled, drifted quietly down the streams. It appeared at sunset, a creature of crepuscular habits, secluded, entirely noiseless in its diving and reappearance in search of food, and making its home in concealed burrows along the river bank. 'On seeing them', one early colonist noted, 'the spectator must remain perfectly stationary, as the slightest movement will cause the timid creature instantly to disappear, so acute are they in sight or hearing'. Wary and reticent, they were very difficult to catch and specimens of *Ornithorhynchus anatinus* were a rarity in the first decades of Australian settlement.

Yet, while direct knowledge on the platypus moved at a snail's pace in the Colony, a burst of theoretical investigation erupted in Europe. Following Blumenbach, the flamboyant French zoological savant, Etienne Geoffroy St-Hilaire, entered the scene. In 1803, working from Home's anatomical observations on both the platypus and echidna, he coined the term 'monotreme' and placed the two animals in a new class, 'monotremata'. Monotreme means literally 'one hole'. This was based on knowledge of the single cloacal chamber, an anatomical feature of both animals. The long struggle to give a taxonomic definition and position to the elusive platypus had begun.

2

THE FRENCHMEN'S GAZE

Nature...leads us a dance in a far continent.

François Péron, *A Voyage of Discovery to the South Hemisphere Performed... During the Years 1801, 1802, 1803, 1804*

While Britain's *Endeavour* voyage opened up rich and unexpected treasures to the scientific world, other nations and scientists hankered to play their part in the unfolding geographic and natural history discoveries and to secure a foothold and reputation for themselves.

The challenge particularly compelled Europe's recognised scientific leader, France. Hence, even while revolution tore at the roots of French society, successive governments— monarchist, revolutionary, Napoleonic—launched a series of elaborately equipped expeditions to visit the new south land and collect specimens for the famous Museum of Natural History in Paris (the Musée d'Histoire Naturelle de Paris).

It was hence to the considerable surprise of the small British enclave of army officers and convicts making a first raw acquaintance with the country at Port Jackson, that two French vessels, *L'Astrolabe* and *La Boussole*, under the command of Comte de la Perouse, sailed into Sydney Harbour in February 1788, a mere three weeks after the British had landed.

Sponsored by Louis XVI, the two ships carried an array of scientists—two astronomers, a naturalist, a botanist, a mineralogist-meteorologist, a geographer, a gardener, botanical and landscape draughtsmen, and two chaplains who were also experienced naturalists. The exploring party spent over two weeks on the northern shores of Botany Bay, which now bears the captain's name, before they sailed out between the great sandstone heads of Port Jackson in early March and vanished into the blue horizon.

In 1791, as tumbrils rolled along the streets of Paris, the French National Assembly despatched Admiral D'Entre-casteaux to search for La Perouse in ships named appropriately *La Récherche* and *L'Esperance*. While the search brought no return—and the wreck of La Perouse's ships would not be discovered off the New Hebrides for nearly 40 years—the expedition conducted cartographic surveys around the Australian coast and, through its outstanding botanist Jacques-Julien de la Billardiere, made impressive collections of Australian flora that yielded two botanical volumes of the voyage and linked Australia to the scientific enterprise of France.

Napoleon Bonaparte's advent to power gave further impetus to French science. On his famous Egyptian campaign of 1798, he carried along with his invading army a corps of no less than 154 savants and a large travelling library in order to investigate everything about the country he planned to command—its ancient monuments, its flora and fauna, the great annually flooding Nile, the medicinal value of mummies and the mysterious birth processes of baby crocodiles.

Science and its national potential ran hotly through the Emperor's veins. But while Napoleon himself was obliged to return to France in 1799 and his Egyptian army succumbed to great loss of life before the united efforts of the British and Turks, many of his 'corps des savants' stayed on to continue their scholarly work and to complete the monumental *Description of Egypt* eventually published in Paris in 24 volumes between 1808 and 1824.

In 1800, the 'little Corsican adventurer' turned his bright eyes southwards. Taking up an agreement with the Institute of France to extend man's knowledge of the habitable regions of the earth, Napoleon commissioned the most expensive and splendidly equipped scientific maritime expedition yet to set out from any country, putting it under the command of the experienced geographer, Captain Nicholas Baudin.

Its scientific and political mission was explicit: to cross the world to New Holland and conduct botanical, zoological, anthropological and meteorological research for the greater glory of France, and, in the midst of war with Britain, to secure from France's own cartographers accurate and strategic hydrographic surveys of the Australian coast.

Few nations showed such flair in naming their exploring vessels as the French. When the smartly titled *Le Géographe* and *Le Naturaliste*, accompanied by the schooner *Casuarina*, sailed from Le Havre in October 1800, they carried on board

23 civilian scientists, including five zoologists and a medley of anthropologists, botanists, mineralogists, physical scientists and natural history artists. Their task, barely more moderate than the Egyptian venture, was the study of the Antipodean physical and natural world and its unknown indigenous inhabitants.

Plying by way of the Cape of Good Hope and Mauritius, the French ships reached Cape Leeuwin on the south coast of Western Australia in early 1801. They then travelled north-wards to Timor and surveyed parts of the indented north Australian coast before turning south to sail the long sweep of eastern Australia to Tasmania.

Later, as *Le Géographe* moved along the southern coast of the huge continent, she met Matthew Flinders' *Investigator* at Encounter Bay. And there off the deserted coast of South Australia, while war raged between their countries on land and sea in Europe, Flinders and his botanist, Robert Brown, climbed aboard *Le Géographe* to meet its captain and members of his scientific crew and drink a toast to the internationalism of science. They would meet again in Port Jackson in 1802 where from June *Le Géographe* enjoyed the friendly hospitality of government and residents for several months.

The French naturalists would meet the platypus in New South Wales. The expedition's zoological recorders, the naturalist François Péron and natural history artist Charles Lesueur, had already made large collections of specimens of faunal life in Tasmania, its offshore islands, and Kangaroo Island off South Australia when they turned their gaze on the fauna around the penal settlement at Sydney.

Trained in medicine and nominated to the expedition as

'zoologist in training' by the celebrated French systematist Jussieu, Péron was alive to European scientific interest in the platypus and to the flurry Australia's improbable fauna had stirred. His descriptive notes and Lesueur's drawing of two living specimens in their natural environment in New South Wales, a brown female and a russet male, were precise. They presented two fully defined platypuses, a sketch of a skull with open jaws and details of the masticatory plates, and— indicative of their grasp of the animal's unusual anatomical organisation—a sketch of the underside of the tail of the female platypus showing the common opening for the genital organs, excrement and urine.

Gathering their specimens for transport to Paris, the two

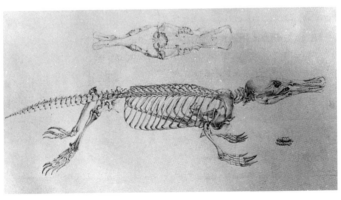

Profile of a skeleton of the Ornithorhynchus, *skull with open jaws and details of masticatory plates. It was the practice of Péron and Lesueur to collect the skeletons of the animals that Lesueur depicted and to take them back to the Museum of Natural History in Paris. This specimen, taken from an animal found in a river near Sydney Cove during the Frenchmen's stay at Port Jackson in 1802, did the round trip back to Australia to appear at the exhibition 'Terre Napoleon' held at the Museum of Sydney in 1999.*

Among the host of Australian fauna which Lesueur painted, two adult wombats, with the young emerging from the mother's pouch, caught his eye on King Island in 1801. Initiated into the natural sciences by Péron on the long voyage to Australia, the youthful Lesueur joined zoological precision with an artist's eye for the singular and engaging fauna of New Holland.

young Frenchmen carefully added the skeleton of a male platypus to their faunal collections.

While Péron described and recorded, and the talented Lesueur shot and painted his specimens, capturing their life's vividness on paper in the swift course of a day, the expedition itself piled up disasters. Dysentery and scurvy pinned it in Sydney Harbour for several months. The tense ill-feeling of the expeditioners assailed the captain. Bad management, poor teamwork and inadequate knowledge or supplies of healthy rations multiplied the expedition's trials. In November 1802, *Le Naturaliste*, under the command of Louis de Freycinet, embarked for France carrying members of the scientific corps who had fallen ill and a great part of the scientific collections including an assortment of live animals.

Nicholas Baudin, extending his time on *Le Géographe* at Port Jackson for several months, would die from tuberculosis at Ile de France in 1803 along with a large number of his officers on the journey home. Death stalked the voyage and only five of the expedition's entire complement of scientists and artists lived to return to France.

The scientific outcome of the expedition, however, was immense. 'Messrs Péron and Lesueur', France's scientific leader Georges Cuvier declared jubilantly, 'will have made known more new creatures than all the travelling naturalists of recent times put together'. There were over 100 000 preserved animals, large and small, of which some 2500 were species new to science. It was a zoological haul significantly surpassing that of the two Pacific voyages of Captain Cook.

The celebrated Geoffroy St-Hilaire was on the quay at Lorient in 1803 to greet the *Le Naturaliste* when she dropped anchor with her preserved specimens and her live Antipodean ark. Three surprised wombats, an emu, a black swan, half a dozen parrots, two dingoes, and a long-necked tortoise landed on French soil.

Baudin had gone to great lengths to transport live animal specimens to the expectant scientists at home. His efforts incurred the open displeasure of the officers whom he turned from their quarters in rough or cold weather to make room for the beasts. When months later a sadly depleted *Le Géographe* limped into Lorient in March 1804, its cargo included one more live emu, a lyrebird and several different kinds of kangaroo.

French science had triumphed in the southern seas. Yet muddle and ineptitude persisted. Péron retreated to the country, riled over the lack of any settlement of his salary. And while the collection of 'shells, fishes, reptiles and zoophytes preserved in alcohol, of quadrupeds and birds stuffed or dissected' (as Péron described them) were unpacked at the Museum of Natural History, it was not until 1804 that he was persuaded to return to Paris by an eminent committee of the Academie des Sciences to write up his brilliant discoveries.

François Péron, quick and independent, held qualities of genius. He was, moreover, the first trained zoologist to land in Australia. Unlike the clever men of great repute—the rising comparative anatomists and scientific theorists of Paris who would grapple with the new specimens of creation from Australia—Péron had actually set foot on the unexamined world of the new fifth continent. He linked keen understanding with direct observation in the field; he sailed across the world with his zoological antennae alert.

Yet once there, observing and describing, he found in the furred aquatic platypus, the pouched kangaroo, the reversely-coloured black swan and the fleet but flightless emu, inversions and contrarieties of nature that challenged a rational explanation.

In the brilliant light of those southern hemisphere days, he was consumed with curiosity and passion. But back in the Old World, he no longer trusted the evidence of his eyes. 'In Science', he wrote, '[New Holland] defies our conclusions from comparisons, mocks our studies, and shakes to their foundations the most firmly established and most universally admitted of our scientific opinions'.

The convict society, with its upside down hierarchies and a population of miscreants who flourished in this New World also surprised him, while the meteorological storms that blew savagely across the settlement at Sydney bore no resemblance

*François Péron, a sharp observer, marvelled at the bizarre
features of both the social and faunal scene he encountered in
the convict colony of New South Wales.*

to the weather he knew in his own land. Nature and society
itself he came to believe 'leads us a dance in a far continent'.

The rigours of the expedition cut short his life. The
talented Péron died in 1810, 'an old man', in the words of
one contemporary, at the age of 35.

After his volume of the voyage was published in 1807, it
fell to Charles Lesueur and the expedition's other surviving
natural history artist, M. Petit, to see the *Atlas* volume of the
Discovery Voyage with its 41 beautiful illustrations, including
the *Ornithorhynchus*, into print.

Yet, despite the expressed pleasure of the blue ribbon committee of the Academie des Sciences—made up of some of the great figures of French science—at the expedition's outstanding results in science, the collections did not become available to researchers. They were placed in the Paris Museum of Natural History, together with Péron's meticulous recording of the external structures, characteristics, and the place and date of acquisition of his zoological specimens, but were not displayed. Péron's great plan for a separate volume on the 'zoography' of the voyage also failed to win the support of the authorities.

Lesueur, too, withheld the results of his labours, declining to place in the Paris museum the drawings he had made to accompany Péron's scientific descriptions. Instead he took them to Philadelphia when he moved there in 1815. It was not until 1838, when interest in the expedition had long since paled, that both Lesueur's drawings and Péron's materials reached the Museum of Natural History at Le Havre, the city from which the expedition embarked.

Working over the materials in the late 1990s, the curator of the exhibition 'Terre Napoleon: Australia through French Eyes', Susan Hunt, suggested that these vast trophies of science were meant to be concealed. They were 'privileged knowledge', and, as such, she wrote, were guarded much as nations jealously guard their seaboard exploration images and satellite maps today. Yet they alone, rather than strategic or political outcomes, were all that remained of Napoleon's ambition to conquer a new scientific empire in the south.

Perversely, it was not the visionary Napoleon, but his Empress, Josephine, who gave celebratory expression to the great venture her husband had so enthusiastically set in train. It was Josephine's private chateau 'Malmaison' outside Paris that received most of the live cargo from the expedition vessels. There, with François Péron's aid, she established a menagerie of Australian animals: kangaroos, emus, parrots, swans, wombats and the tortoise, to roam in grounds that also included the seeds and cuttings from Australia's acacia, casuarina and grevillea trees. The motif of the black swan adorned the Empress's bed.

Yet the wheel of history swings full circle. In the last two decades of the twentieth century, the detailed story of the forgotten voyage has been disinterred, the art of the Baudin expedition published, and drawings, maps, journals and objects of a rich and distant undertaking brought to public view.

In one of those turns of the wheel that reward the historian and bring the researcher keen delight, I found myself in 1999 standing at the exhibition 'Terre Napoleon' which had travelled to Australia for display in Sydney and later at the National Library of Australia in Canberra. And there, protected yet vivid under its shiny plastic frame, was the skeleton of the male platypus *Ornithorhynchus anatinus* caught by Péron in 1802 in New South Wales and returned nearly two centuries later to declare its part in the story of a great scientific enterprise. 'The skeleton', informed the inscription, 'is one of the only specimens which is positively provenanced to the Baudin voyage existing in the Musée d'Histoire Naturelle de Paris'.

Breaking on French eyes in Lesueur's striking drawing, published in 1811 in the *Atlas* volume of the voyage, the platypus was, indeed, a strange creature who spoke to a public exhilarated by distant exploration and discovery. But to the comparative anatomists puzzling on its place in the scientific register, although confirmed in form, it was as scientifically enigmatic as before—a taxonomic riddle from across the globe.

MARSHALLING
THE ANIMALS

It is a great misfortune to science that zoological
systems are necessarily not merely the creatures of
human invention, but, to an extent also, of human
fancy and caprice...the best zoological arrangement
will ever be capable of improvement.

Baron Cuvier, *The Animal Kingdom*, 1817

It would appear that everything that can be, is.

Comte de Buffon, *Discourse sur la maniere
d'etudie et de traite l'histoire naturelle*, 1749

Naming the animals was man's first task in the Garden
of Eden. Second only to naming was the engrossing but
infinitely more demanding business of arranging and classifying
the birds and beasts of the field and all the insects and creatures of land and sea to form some system of order and
relationship. The discovery, naming and classification of species

became for centuries a central theme and enterprise of natural history.

The key to any system of classification—and to an understanding of the natural world—was the organisation of groups, subgroups and other relevant divisions by prescriptions that recognised their commonalities and marked their divergences. Aristotle early set his mind to the question with his *Historia Animallium* in 334 BC. Gazing upon the sweep of creatures known in classical times—predominantly those of the Greek mainland, its islands and colonies and some drawn from menageries and travellers' tales—the observant philosopher laid out a map of the animal kingdom according to the ways in which animals 'were like to and different from each other'.

Faced with a Herculean task, he divided his cavalcade into two major groups, those with blood, and the 'bloodless' invertebrates. He then went on to describe them according to their form and structure, their manner of operation, locomotion, habits and food. What Aristotle sought in the variety of animal structures confronting him were commonalities of function, parts with the same kind of attributes, and correlations between the parts. He separated out as a group the quadrupeds who shared the physical characteristics of four feet, fur and blood, who were terrestrial, and gave birth to live young.

Pushing new additions into his filing cabinet of creatures which he gleaned from fishermen, hunters, farmers and his own acute observations and dissections, he turned his eye upon the animals' organs. He noted the position of the uterus, observed the placental structure, differentiated the 'viviparous' quadrupeds by the perfection of the young at birth, and separated the 'oviparous' fishes, who laid their eggs for external hatching, from the fishes, sharks, vipers and dogfish who were 'ovoviviparous' in kind and hatched their young from eggs inside the body.

With a cast of 80 mammals, 180 birds, a cluster of marine animals, 130 fishes and an array of insects, Aristotle's treatise was an outstanding descriptive zoology of ancient times and the first scientific classification of living creatures.

Throughout the Middle Ages and well into the seventeenth century, his descriptive catalogue stood. Adopted by collectors and would-be naturalists arranging their cabinets of curiosities and poring over their special collecting fads, it enabled them to impart some sense of order and relationship to the diversity of nature's forms. As late as the mid-nineteenth century, the comparative anatomist Richard Owen heaped praise on his classical precursor, declaring with perhaps exaggerated zeal that in the two thousand years since Aristotle wrote, 'the ideas of learned men regarding nature and classification of Mammalia received no improvement'.

There were other prime contenders. The late-seventeenth-century systematist John Ray, whose very Englishness secured him pride of place in the advance of natural history, adopted the same classificatory boundaries as Aristotle when he came to produce his *Synopsis Quadrupedum* in 1693, though he offered some alternative subdivisions. From his watchful knowledge of the organic world, Ray selected 'classes' and 'orders' from characteristics that linked groups of animals through some significant feature. For land vertebrates like mammals, birds and reptiles, he chose the method of reproduction as an important marker, but he also added questions of blood circulation, respiration and locomotion to emphasise differences. Thereafter all mammals were expected to be viviparous and the presence of 'mammae', by which females nourished their young, signalled for Ray and his successors the difference between mammals and all others. In time, (though the old term lingered long in zoological writing) 'mammal' came to replace the original 'quadruped'.

Ray stood as the father of systematic natural history, a taxonomist who sought to define a system of organic nature through a clear formulation of identifying taxa. His system for assigning defined categories in natural history was said to be 'at once the shortest and most comprehensive' and his volume found a welcome place in every naturalist's library. Ray also infused his guiding format with natural theology. Everything in nature, he affirmed, proceeded from a Creator whose eye and hand encompassed every archetype known to man or yet to be revealed. His effort crested a wave. Since 1500, as adventurous navigators began to crisscross the oceans, the number of registered species had doubled from 150 to 300, and pressure for identification and classification of these discoveries put increasing demands on both the exactness and the flexibility of taxonomic systems.

Into this arena of definition and classification, this international attempt to give a scholarly framework to natural science, stepped the charismatic and idiosyncratic French naturalist and philosopher, Georges Louis Leclerc Comte de Buffon, director of the Jardin du Roi in Paris. At just under five feet tall, Buffon nevertheless strode his world like a colossus. Natural history in all its aspects: zoology, physiology, mineralogy, geology, biogeography, anthropology poured from his pen, and when he died in 1788, at the age of 80, he had published 36 great tomes of a *Histoire Naturelle*. His volumes became best-sellers and his lively and opinionated style scooped up a readership and influence far greater than that of any other naturalist of his day.

Buffon eschewed arrangement. Aware of the disorder in nature, he distrusted classifiers who sought to imprison nature into an artificial system. Rather, he filled his volumes with detailed, superbly crafted, individual descriptions of his bestiary (each group or type encased within its chapter) which brought his procession of birds and animals vividly to life. His writings blazed with personal observations and impressions that engaged the reader and left little doubt of the author's view on any given subject. Hyperactive himself, he most despised that early zoological arrival in Europe from South America, the sloth.

This animal's defects—he ticked them off conclusively—included:

> no incisor or canine teeth, small and covered eyes, a thick and heavy jaw, flattened hair that looks like dried grass ... legs too short, badly terminated ... no separately movable digits, but two or three excessively long nails ... Slowness, stupidity, neglect of its own body, and even habitual sadness, result from this bizarre and neglected conformation ... These sloths are the lowest form of existence in the order of animals with flesh and blood; one more defect and they could not have existed.

'Le style', he asserted, 'est l'homme même' (The style makes the man). No systematist of the accepted kind, Buffon took

An illustration from Buffon's Natural History. *Buffon grouped these four animals—the Great Ternat Bat, the Tamanoir, the Pangolin and the Armadillo—together in his natural history system according to their function and activity since, as he put it with engaging simplicity, the last three 'had many things in common; they fed on ants, sucked honey, and were tamed and domesticated easily'. 'They go so slowly', he added, 'that a man can overtake them'.*

species as his touchstone and saw them as more important than the genus, the order or the class; more flexible in adaptation and capable of breeding across species. Nature, he believed, 'amazes more by her exceptions than by her laws'. But he also perceived a unity of plan in nature and a community of origin from one single form. Starting from the top down, from the highest orders, he got no further in his voluminous descriptions of the animal kingdom than mammals and birds, and the task of completing the reptiles and fishes was left after his death to his colleague, Lacépède. But his lively perception and information offered a new perspective on natural history while his language lingered in such classic maxims as 'Cet animal is bien méchant. Quand on l'attaque, it se défend'. (This animal is very bad. When it is attacked, it defends itself.)

Yet Buffon, that 'greatest enemy of Arrangement' as British naturalists dubbed him, was, despite his great popularity in his lifetime, to remain less influential in the long run than his arch-rival and close contemporary, Carl Linnaeus. Born in the same year and immersed in the same large questions, the two men were at opposite ends of the classificatory pole. The Swedish Linnaeus, fixing his sights on genera and species described again in the Latin tongue, imposed a strict taxonomic order on nature that overturned the old systems while absorbing their basic aspects within his own. The Linnean system of binomial nomenclature (having two descriptive names) made it possible to assign each plant and animal its own unique place in a comprehensive system and to encompass large and small categories within its hierarchies. Thus the species of the domestic dog (*Canis familiaris*) and the wolf (*Canis lupus*) of the genus *Canis* fell neatly within the family Canidae, itself within the order Carnivores. Carnivores, in turn, belonged to the mammal class, while the class was contained within the phylum of vertebrates.

Like Ray, Linnaeus also attached importance to the functional implications of reproduction. He made the presence of mammae (milk-secreting glands) and the suckling relationship of mother and young as the defining criterion for the class of warm-blooded animals which he named 'Mammalia'. His warm-blooded 'quadrupeds'—quadrupeds with a four-chambered heart and double circulation—were defined categorically by Linnaeus as viviparous—bringing forth their young alive—and mammiferous. His *Systema Naturae*, first published in 1735 with descriptions of 549 animals, ranged across nearly 6000 species in its last edition.

While Linnaeus had, as one British naturalist reflected tartly, superseded the old forms, 'only to exalt his own name', the Linnean system came to dominate natural history. Banks honoured the Swedish naturalist as 'our Master', while *Systema Naturae* was repeatedly republished, translated and expanded. In East Anglia, an Anglican clergyman grappling with a taxonomic epic, sang Linnaeus' praise in verse.

> And thus not only were arranged, and classed,
> The subjects of my vegetable world,
> But every *Beast*, and every *Bird*,
> The amphibious tribes, tenants of the
> Congregating waters of the deep, were
> Summonsed all, and all again displayed in order
> To receive from thee their names...

38

There were, of course, recurring discontents. Naturalists fretted over their collections and systems, consensus was rare, animals such as the elephant provoked persistent debate about its placement within the scheme, and many felt free to use their own taxonomic discretion. The distinguished late-eighteenth-century naturalist John Hunter (whose private collections would be acquired by the Royal College of Surgeons to found the Hunterian Museum) grouped his specimens of comparative anatomy, including a good spread of mammals and marsupials, according to their internal organs. Yet despite many niggling disagreements, mammals and quadrupeds remained the most stable classifications in the zoological register.

Aristotle, and specifically Ray and Linnaeus, had marked the critical distinction between viviparous and oviparous (egg-laying) animals. All mammals were expected to give birth to live young. 'The main character which distinguishes [them] from all other animals', the German Blumenbach underlined the point firmly in his *Manuelle d'histoire naturelle* in 1803, 'are the mammae with which the females nourish their young'. The exploratory expeditions of the eighteenth century, fanning out about Mediterranean shores and around North America and the Pacific, had already dropped a great array of curious new creatures on the specimen tables of British and European naturalists. Yet this influx had not challenged the commonly accepted divisions between the major vertebrate classes—mammals, fish, birds and reptiles.

Into this arena of categorisation and allotment came the kangaroo, the echidna and the platypus. The kangaroo was readily identifiable as a marsupial mammal which bore and suckled its live young, although the question as to how the pea-like baby got from the uterus to the teats in the pouch would puzzle zoologists until 1954. But the platypus?

The surprised Dr George Shaw, confessing ignorance of the amphibious animal's real nature, placed it in the lowest Linnean order of Bruta, next to the anteater, and his classification was adopted by the British naturalist William Turton who integrated it into the Linnean system in his edition of Linnaeus' *A General System of Nature* published in London in 1802. In Göttinberg, Blumenbach after his first meeting with the dried platypus specimen also consigned it to the family of anteaters, sloths and armadillos. He judged his *Ornithorhynchus* to be like that other anomalous creature, the bat, a likely transitional form between mammals and birds.

But the platypus specimens preserved in spirit presented naturalists with a different scenario. The internal organs suggested that the animal resembled a lizard more closely than it did a bird or mammal. It appeared to have no mammary glands, and this, on traditional systematic reckoning, flashed the signal that it did not give birth to live young and must then be oviparous.

Everard Home, in England, was also forming his own conclusions. Dissecting the platypus in 1802 for a paper on its anatomy—a paper that would furnish the foundation description of the platypus's anatomy for the next 25 years— he noted that the structure of the ear and shoulder girdle combined mammalian and reptilian features and that the internal organs, using only the one chamber (the cloaca) for reproductive and excretory functions, linked its reproductive system to that of reptiles and birds. The absence of a well-formed uterus and the seeming absence of nipples persuaded Home that the Antipodean 'duck-bill' was most closely analogous to ovoviviparous reptiles which produce eggs that hatch within the body of the mother.

The different interpretations begged the question: each naturalist sought to shoehorn the little animal into their different prescriptive forms. Each sought to accommodate it within fixed and long established categories. But in each perspective, the warm-blooded, furred, genitally peculiar animal from the Antipodes plainly represented some kind of transitional form, an unexpected bridge between the categories of mammal/quadruped and reptiles and birds.

No animal, indeed, was to rub more strenuously up against the prevailing taxonomic categories than the paradoxical platypus. It cut across recognised boundaries and its failure to fit the classificatory prescriptions underlined its oddity and uniqueness. Preparing his forthcoming course in natural history in 1814, Busick Harwood, professor of Anatomy at Cambridge University, promised to devote special attention 'to the astonishing union of the characteristic distinctions of all the Classes in that extraordinary animal'. Bizarre, elusive, and in short supply for examination by comparative anatomists, *Ornithorhynchus anatinus* was to hold nineteenth-century naturalists in perplexed and persistent thrall.

In the struggle to clarify, patriotic impulses and scientific kudos ran side by side. As the British and French expeditions proved, scientists were particularly anxious to uphold the right to classify specimens brought back from their own country's outposts and exploratory voyages. It was, perhaps, in this spirit of national one-upmanship—as well as an expression of the dictum that 'science was never at war'—that Sir Joseph Banks in 1802, when the French expedition was itself pursuing the natural history of New Holland for the glory of France, reportedly despatched a specimen of the platypus to Napoleon.

The cosmopolitan Geoffroy St-Hilaire highlighted another aspect of the mood of national and European ownership of nomenclature and taxonomy when he reflected on prevailing discourse about the kangaroo. 'In Europe', he wrote, 'where our opinions are regulated, *a priori*, by what happens ceaselessly under our eyes, and where, in this respect, our theories on generation have been somewhat fixed, we have profited by a certain vagueness which prevails in the observations which travellers have reported to us on the subject of marsupials, in order, dissimulating some facts and exaggerating other circumstances, to restore the mode of generation of these animals to a common standard'.

In addition to an overriding need to provide clear and defining zoological criteria and hierarchies of classes, genera and species into which new discoveries could be pressed, the ancient concept of a 'Chain of Being' was also fundamental to the taxonomic view.

Aristotle first voiced the notion of a *scala naturae* that arranged organisms in a linked progression through an increasing complexity of structures from the lowest to the highest form. Embedded in theological thinking across centuries, however, the Scale of Nature asserted the immutability of species: God, it was believed, had created the entire chain when he brought the universe out of chaos. But the concept of the chain underwent change.

Buffon, in his inimitable prose, embraced the idea of gradation of species. Everything, with its infinite variety, figured in a unity of plan and descended from one advanced single form. The indomitable Linnaeus, God-like himself in his self-image, firmly upheld the fixity of species and the one act of Divine Creation, and cherished the thought that his own comprehensive botanical and zoological system gave him 'a glimpse into the secret cabinet of God'. But Linnaeus also proclaimed a lineal 'Chain of Being' and was the first to link the chain from apes to man. For Linnaeus, man was unique in the chain, exclusively empowered with intelligence to study the works of the Creator and in his dignity exalted above all other animate creatures.

The concept of the chain also found strong roots in the prevailing teleological belief in 'separate creation'. While the links, affinities and ascensions in God's creative handiwork could be perceived and acknowledged, it was widely believed that the Deity in his infinite wisdom had made his individual species, step by step, in distinct separately creative acts.

By the early nineteenth century acceptance of the 'Chain of Being' underlay the very structure of zoological thinking and classification. Yet as large collections of new animal species from distant places assembled in cabinets and museums, the chain showed ominous signs of tangling and of offering some ill-assorted and confining links. Well before an uneasy Darwin scribbled in his notebook that 'all animals have never at any one time formed a chain' other naturalists were noticing stress points and insisting that the links and affinities seen between species and families must not be allowed to become restrictive fetters.

In addition to the platypus and echidna, the Australian marsupials in general raised questions and rattled the links of the chain. At first when the 'wonderful animal', the kangaroo, reached Britain, naturalists tried to place it within one of the established mammalian orders, classing it by way of its pouch, with the opossum. The opossum was a veteran on the European scene. Discovered in South America in 1500, a specimen had been brought back alive and presented to Isabella, Queen of Spain, where its strange frontal pouch delighted observers and impressed naturalists for the special protection it gave the animal's young. Several sixteenth-century writers examining the first non-royal specimens observed that once the young were born, the mother trans-ported them in her external belly 'somewhat like a bag'. Several others who had the good fortune to visit the New World reported that the opossum young were delivered from the uterus and entered the mother's pouch after birth. She became their perambulator and their cradle.

But it soon became clear to the examiners and dissectors that the opossum's reproductive system was unlike that of any other mammals. Several perpetuated the idea that the pouch was the place of conception since the young found there were minuscule and poorly formed. The newborn opossum, indeed, weighed in at no more than 2.5 grains (1 grain equals 0.0648 grams). Twenty, if so tested, would fit on a teaspoon. In contrast to placental mammals, the opossum uterus was not developed for the nutritional, respiratory and excretory activities that occur while the foetus develops. Room for development in the marsupial uterus is extremely restricted. When expelled from the uterus, marsupial young are still embryonic; they are much less developed than placental offspring and face an extended period of development attached to their mother's teat in the pouch.

Such insights came slowly. Over a century and a half after the discovery of the opossum, writers still clung to versions of the pouch birth. In 1629, the Dutch voyager Francis Pelsaert, wrecked with his vessel on Houtman Abrolhos off the Western Australian coast, recorded in his journal the first account of an Australian marsupial, a species of the Tammar wallaby (*Macropus engenii*), which he called 'cats'. Their manner of generation, he wrote excitedly, was

> exceedingly strange and highly worth observing. Below the belly the female carries a pouch, into which you may put your hand. Inside the pouch are her nipples, and we have found that the young ones grow up in this pouch with the nipples in their mouth . . . It seems certain that they grow there out of the nipple of the mammae, from which they draw their food, until they are grown up and able to walk.

An anomaly of this kind was challenging. When John Ray came to include a description of the opossum into his systematic work, he was oppressed by a conflict between accepted knowledge of the birth process of 'normal' viviparous animals and the available information that the opossum was distinguished from all other mammals 'by a singular and certainly admirable pouch or open uterus in which the young are received after birth'. An *external* uterus? Discomfited, Ray insisted that the anomalous reproductive process be confirmed by repeated observation in the field.

It was the redoubtable Buffon, however, who fortified by new data on opossum organs and by evidence from the British naturalist Edward Tyson, resolved the conundrum of opossum birth. Tyson had secured a dead specimen of a female animal from the zoo and, finding an internal uterus (though very different from that of placental mammals) deduced that the pouch was used solely for the protection of the young. With keen insight, Buffon went one step further and explained marsupial reproduction as a 'two-stage' gestation in which part of the foetal development occurred in the uterus and part in the pouch.

True, Buffon's explanation assumed the opossum's birth was premature, an error that gave rise to much speculation, hot air and interminable papers on how an embryo incapable of independent action got from the mother's vagina to her teat. But his two-stage model marked an anchor-point for the biological theorists.

The arrival in Europe of the kangaroo and other Antipodean marsupials opened the debate anew. Together with the platypus and echidna, the Australian marsupials presented biologists with rich new organic forms that both excited the imagination and sparked abundant questions concerning anatomical anomalies, the Chain of Being, the very basis of classificatory systems and the geographical distribution of species. Here was an unanticipated fauna disturbing the systems that, hitherto, had served effectively to marshal the animals.

Slowly assimilated, and fought out through personal and national rivalries, the evidence from the Australian monotremes and marsupials would play a dynamic part in formulating a new theoretical framework for biology.

4

THE WRANGLING
SCIENTISTS

Even among naturalists there was room not only
for scholarly debate but for deeply personal or
eccentric alternative interpretations.

Harriet Ritvo, *The Platypus and the Mermaid*

Since his famous paper on platypus anatomy, read to the
Royal Society in 1802, Sir Everard Home was in the zoological ascendant, the leading British anatomist of his time.
Yet, while British interpretations on the anatomy of the
strange animal paused, France's leading experts were staking
strong and divergent claims.

The influential Geoffroy St-Hilaire was an early protagonist
in the debate. Geoffroy (who had added his given name to
his surname in his youth and was known thereafter by it alone)
was special. A dashing and brilliant figure in French biological
science, he was a polymath and a romantic. At fifteen he had
become a canon in the church with prospects for a religious
career, but with the outbreak of the Revolution he turned to

law, then medicine, and embraced revolutionary ideas. At the beginning of the Terror he was responsible for the release of Abbé Hauy—one of the great founders of crystallography, imprisoned because he was a priest—and, in recognition of his efforts, Geoffroy was appointed professor of quadrupeds, birds, reptiles and fish at the Paris Museum of Natural History in 1793 at the age of 21.

When Napoleon organised his famous Egyptian campaign, Geoffroy enthusiastically joined and from 1798 to 1801 his scientific enquiries in Egypt were filled with adventure and danger. He successfully rescued the important natural history collections he had made while travelling up the Nile to Aswan, and on his return to the museum in Paris devoted himself to descriptive zoology and classification, resuming the research on marsupials that he had earlier begun.

The platypus and the echidna quickly claimed Geoffroy's attention and, working from Home's descriptions, in a paper of 1803 he grouped the two in the new taxon which he named Monotremata, describing it as characterised by animals 'digits clawed; having no true teeth; a common cloaca opening to the exterior by a single orifice' and an absence of mammae. But Geoffroy did not indicate precisely where this new order should be placed in the classificatory system. The male specimen of the platypus brought back to the Museum of Natural History by Péron later that year—the first specimen of the paradoxical animal to reach French researchers— evidently did not (possibly because of its gender) add anything to lure him further into print.

During 1809—some six years after Geoffroy—Jean Baptiste Lamarck entered the fray. Destined to become a major and controversial figure in the evolutionary debate, Lamarck had early spread his interests over botany and zoology and, under the powerful patronage of Buffon, entered the Jardin du Roi in Paris. Five years later, when that institution reformed to become the Museum of Natural History in 1793, he was appointed in charge of its division of invertebrates, where he came to be known as the 'Professor of Insects and Worms'. In his famous *Philosophie Zoologique*, he created a new and different class for the platypus and echidna. In his opinion, the two animals were not mammals since, on the evidence available, they had no mammary glands. Neither were they reptiles, since they possessed the mammalian characteristic of the four-chambered heart. Nor were they birds as their lungs differed from those of avian species and they had no wings. Lamarck named the new class *Prototheria*, a name which remains in existence today.

A young French doctoral student, Henri Marie Ducrotay de Blainville, went so far as to devote his doctoral thesis to the place the *Ornithorhynchus* and echidna should occupy in the natural system. Fresh and uninhibited by old ideas as good students are, Blainville contended in 1812 that 'too much had been written on too little direct observation' on the subject of these ambiguous animals. Not only were there few specimens from which to draw conclusions—and these seldom well preserved—but, as he remarked 'zoologists, one after another misdirected by badly interpreted words, have finally come to the point of forming a distinct and separate class which they assume to be intermediary between mammals, birds and reptiles'.

Blainville, who was destined to rise high as a comparative

anatomist and independent thinker, saw mammals as descending with decreasing complexity of structure and organisation from the primates down through the marsupials to the monotremes. He divined many resemblances between the platypus and echidna and the marsupials, a point he was the first to make, and he judged that while there was an apparent absence of mammary glands, the monotremes belonged regardless in the class of milk-giving mammals. He assigned them to a separate order, *Ornithodelphia*.

No one held a more illustrious position in French biological science than Georges Cuvier. From the last decade of the eighteenth century, he had swept like a meteor across the zoological sky. Initially a candidate for the church, the young Cuvier had become entranced by natural history during a six-year residence on the Normandy coast until he was brought through Geoffroy's influence to the post of assistant professor of Comparative Anatomy at the Museum of Natural History in Paris. Three years Geoffroy's senior, Cuvier did not join Napoleon's Egyptian 'corps des savants'; remaining instead in Paris, he worked on the museum's invaluable specimens of mammals and birds.

Indefatigable and committed to a comparative anatomy based on organic structures and their relationships from one taxon to another, he published his *Tableau elementaire de l'histoire naturelle des animaux* in 1797, classified the entire collections of the museum in 1804 and, turning to his famous palaeontological reconstructions of fossil bones in the Paris Basin, extended his interpretations of nature to extinct forms.

Cuvier was an example of the dictum from the Gospel according to St Matthew, 'to him that hath shall be given'.

Beginning his scientific career humbly in Normandy as a very thin and poor young man making excellent drawings of dissected invertebrates and fish, he rose swiftly, reaping praise, honours and great wealth, to became a very fat, splendidly robed and famous man rejoicing in the nickname 'Mammoth'.

In his mature work *Le Règne Animal* published in 1817, Cuvier placed the platypus and the echidna as a third genus in the order of the toothless Edentata (anteaters and sloths). But he pressed the point that 'since travellers believe that these animals lay eggs', it was desirable 'that a trained anatomist describe exactly these eggs, their origin within the body and their development after being deposited outside the body'. It was surely an examination, this pre-eminent figure in world zoology suggested, that could be made by the many medical men who visited Port Jackson.

The French researchers, in true Gallic fashion, differed animatedly on the animals' taxonomic position and the mode by which they produced their young. Geoffroy and Lamarck believed they laid eggs and were oviparous, since in both their views, the animals were not mammals. Blainville assumed a viviparous birth with live young. Cuvier, while consigning the monotremes with the amphibians (which at that time did not constitute a class by themselves) in the order of Edentates, recognised the peculiarity of their generative structure and awaited further evidence as to their viviparous or oviparous birth.

There the matter stood by the early 1820s as second-hand reports of an egg-laying platypus trickled in from distant New South Wales.

The year 1824 would mark a crucial turning point. In

that year the German comparative anatomist Johann Meckel, whose watchful interest in the platypus and echidna dated back some fifteen years, published a paper in Leipzig announcing that he had found mammary glands in the platypus.

True, the glands appeared primitive, they opened directly onto the skin and there was no sign of nipples. But mammals the animals were, he declared. Their offspring might well be born live, although Meckel advanced the important and then revolutionary view that lactation did not necessarily require a live birth. Marsupials, for example, brought forth live young in very embryonic form. Monotremes, he suggested might well be 'egg-laying mammals'. As such, they would undoubtedly represent a transitional form.

Personal interpretations of Meckel's discovery now sprouted like mushrooms in a freshened field. Blainville eagerly accepted the German anatomist's announcement since his proposition encompassed Blainville's own chosen outcome that these mammals might produce live young. The great Geoffroy, however, offered a hot rebuttal.

Since he had believed in the oviparity of the platypus for some 25 years, Geoffroy rejected Meckel's discovery of mammary glands. They could not be mammary glands, his paper of 1826 asserted, because in the case of the platypus, the very absence of nipples would make feeding especially difficult for an animal whose beak was so ill-suited to sucking. The so-called glands were much more likely to be structures resembling the lubricating glands of the salamander.

Satisfied with his own interpretation, Geoffroy claimed that the animal's anomalous organic structures meant that the platypus was a mammal, in effect, but without the true mammalian characters. It was this, he claimed, that justified his original creation of Monotremata as a special taxonomic order.

Meckel fought back, cheered that the formidable Geoffroy had at least confirmed the existence of the disputed glands even while declining to recognise them as 'mammae'. Meckel was confident of his own discovery. In his more detailed and finely illustrated publication on the *Ornithorhynchus* in 1826, he declared that while platypus mammary glands clearly differed structurally from those of other mammals, their function was similar. From the specimens at his disposal, he had determined that the glands were quite large in older females but almost completely lacking in younger ones.

The authoritative Cuvier, temporising wisely, writing his overview of the progress of the natural sciences in 1827, again stressed that this was a discussion that could 'hardly be settled except by those who observe the living animal'.

The German anatomist Karl von Baer lent his support. Eager to assist his colleague and examine his dissections, Baer testified that Meckel's discoveries were indeed mammary glands. However, on the question of generation, there was merit, he thought, in Cuvier's plea. The question of viviparity or oviparity could not be resolved unless the animal was seen to emerge from the egg.

Geoffroy, having since had an opportunity of dissecting a female platypus, again fiercely rebutted Meckel. His view turned on the point that the contentious mammary glands possessed none of the characters of a true mammary gland, and there was no trace of nipples. He hence decided that the platypus glands were similar to the glands for secreting lubricating liquid disposed along the flanks of fishes and aquatic reptiles. Q.E.D.

The academic papers of the protagonists, each published in their own language in their respective national scientific journals, proclaimed the international character and span of science. Naturalists communicated across linguistic and national boundaries. But each, even while limited by lack of access to well-preserved specimens of the animal in dispute, fought tenaciously for his own 'scientific truth'.

In England, Sir Everard Home, for long the arch-authority on platypus anatomy, was particularly threatened by these events. Meckel's detailed examinations and papers challenged his reputation as a leading comparative anatomist and the position he had staked out from his own dissections that the platypus produced its young ovoviviparously, from eggs hatched within the body of the mother.

No animal, as the President of the Royal Society of London reflected in something of an understatement in his annual address to that prestigious gathering in 1831, 'has ever excited the curiosity of naturalists more than the platypus'.

A restatement of Britain's position in this rising conflict was now urgently required. In 1828, Richard Owen, a recent appointee to the Royal College of Surgeons assisting in the cataloguing of the collections of the Hunterian Museum, entered the debate. With youth, pluck and driving ambition on his side, Owen prepared to defend Home's legendary status and British supremacy in this zoological affair.

Owen had early placed himself on a swiftly moving escalator to a pre-eminent career. Losing his father at the age of five, he quickly entered a man's world, acquired a strong scientific leaning at school, and joined the Medical School of Edinburgh University and later St Bartholomew's Hospital in London. Here, in brisk stages, he picked up a considerable knowledge of comparative anatomy and dissection and, while privately practising medicine, was appointed to the Hunterian Museum as assistant conservator in 1828 at the age of 24.

Challenged, he rustled up an old female platypus specimen from the museum's collections and took the first step in a field of conflict in which he would play a commanding role. His initial sally was not assisted by the state of the elderly specimen, and his dissections have not survived. But Owen's enterprise led to appeals being sent soon after to the private secretary of the Governor of New South Wales to procure specimens of impregnated female *Ornithorhynchus* for Sir Everard Home.

Home himself, his reputation endangered, was soon back at the dissecting table although the enquiries made by the Governor's assiduous secretary failed to winkle out much in the way of materials. 'The female when it is in an impregnated state', one correspondent replied to the secretary's overture, 'is very shy and more difficult to be procured than ever'. But another settler, Thomas Busby, caught three female platypuses which he preserved in spirits and, in 1831, consigned to England.

The problems of efficient faunal collection were substantial. Specimens in prime condition called for careful techniques. Preservation by bottling or placing in a 'keg' required one deep immersion of the specimen in spirits, and a subsequent re-immersion with fresh spirits to achieve satisfactory results.

However, with five specimens newly acquired from the Colonies before him, Home brought in his one-time assistant, Francis Bauer, brother of the illustrious *Investigator* artist, Ferdinand Bauer, and now a botanist at Kew. Working on three acceptably preserved animals Bauer after 'most minutely examining the specimens' assured Home triumphantly that he 'could not find the slightest trace of anything like mammilary glands' and 'found everything precisely the same as you have given it long ago'.

The findings endorsed Home's own dissection of one specimen and the distinguished anatomist assured the Royal Society that the animal did not possess the slightest appearance of mammae. Meckel had erred, either, said Home, because the parts of his specimen had been kept too long in spirits or—the shaft went home—because the German scientist's findings were magnified by 'the imagination of a mind prepossessed with the existence of Mammae'. However, 30 years after his pioneering work on the subject, Home signalled a major change of mind. Abandoning ovoviviparity, the illustrious old anatomist declared that the generative process of the platypus was probably, like non-lactating egg-laying reptiles, oviparous.

Into this hotbed of controversy and theoretical speculation came a letter from New South Wales. A reputable reporter, the Hon. Lieutenant Lauderdale Maule of the 39th Regiment stationed in the Colony, imparted information to the Zoological Society of London which conveyed tantalising reference to suckling platypuses and platypus eggs. 'During the spring of 1831', wrote Maule, 'being detached in the interior of New South Wales, I was at some pains to discover the truths of the generally accepted belief, namely, that the female Platypus lays eggs and suckles its young'. Several of their nests, he recorded, were discovered with considerable labour and difficulty. 'No eggs were found in a perfect state, but pieces of substance resembling egg-shell were picked out of the debris of the nest.' In one nest Maule found a female and two young. He kept the female alive for two weeks by feeding her on worms and when she died, 'on skinning her while yet warm', he noted, 'it was observed that milk oozed through the fur on her stomach', although no teats were visible. But, on proceeding, 'two teats or canals were discovered, both of which contained milk'. The body of this female was forwarded to Richard Owen.

To this information, the inimitable Geoffroy countered with lively French pragmatism: 'If those glands produce milk, let's see the butter!'

As the papers of the contending European scientists ricocheted from their national journals, Owen was drawn directly into the controversy. In June 1832 he launched his major paper 'On the Mammary Glands of the *Ornithorhynchus paradoxus*' which was read, according to custom, by a colleague and Fellow, before the Royal Society of London. With characteristic clarity, Owen set down the puzzling problem of the Australian monotremes:

> The extraordinary nature of the monotrematous quadrupeds of Australia cannot be illustrated more forcibly than by observing that it is still doubtful to what class of animals they properly belong. In the confines of the animal kingdom ...it is less surprising that a species should occasionally

be discovered, either so devoid of external character, or of
a form so strange, as to occasion a difficulty in ascertaining
its class...But the difficulty occurring with respect to a
hairy quadruped, affords one of the most unexpected, as
well as interesting problems in natural history, and renders
acceptable the smallest addition to the series of facts already
ascertained respecting so anomalous a creature.

Carefully examining five fully-grown female platypuses
including one sent by Maule, he noted the glands' appropriate
position in the abdomen of the females and their absence in
males, and gave unequivocal confirmation of the existence of
mammary glands. With marked sophistication, he compared
the size of the ovaries with the lactating parts and found the
glands enlarging in the different female specimens 'as the ovary
appeared to have recently executed its function'.

A young man rising, Owen's paper contradicted both the
great British scientist Home and the formidable Geoffroy on
anatomical grounds. The following year, in a paper on the
Echidna hystrix, Owen was able to demonstrate that the echidna,
the only other monotreme, possessed the same mammary
apparatus as the platypus.

Carrying Meckel's conclusion further, Owen firmly demonstrated that the monotremes should be classed with the
milk-nourishing Mammalia. They did not require a separate
order. But, he allowed, this proof did not resolve the question
of their reproductive process. The discovery, indeed, he wrote
'leaves us as much in the dark as we were before respecting
its [the platypus] mode of generation, and equally dependent
on the exertions of those naturalists who may have the good

fortune of observing facts in the living animal respecting this most interesting and important subject'.

The critical answer to the monotreme question lay hidden in the Australian bush.

5

THE LAND OF
CONTRARIETIES

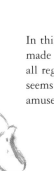

In this remote part of the earth, Nature (having
made horses, oxen, ducks, geese, oaks, elms, and
all regular productions for the rest of the world)
seems determined to have a bit of play, and to
amuse herself as she pleases.

Reverend Sydney Smith, Sydney
1819, in Bernard Smith, *European
Vision and the South Pacific
1768–1850*

The problem of locating specimens to settle the rising
debate on the platypus caused maddening delays for
British and European men of science. Yet, since the animal's
first disconcerting appearance in England in 1799, materials
and information from New South Wales itself was fragmentary
and contradictory.

Joseph Banks' overtures to his colonial and botanical
contacts during the new century's first decade had managed
to extract remarkably little new information for naturalists

at home about either the platypus or echidna. As late as 1818, two years before his death, the Royal Society's ageing but indefatigable president was still applying gentle pressure to colonial governors to turn the vice-regal mind to the mysteries of the 'duck-bill's' reproduction.

The study of zoology took root very slowly in Australia. For the first decade or so after the establishment of the small penal settlement at Port Jackson on Sydney Cove in 1788, there was a burst of interest in natural science. Governors, ships' surgeons and officers of the First Fleet—educated men, often with artistic leanings—recorded their impressions of the extraordinary animals in a state of enlivened wonderment at what they saw. With facile brush and pen, they captured the small jewel-like birds, the laughing kookaburra, the rainbow-hued parrots, the aberrant black and white cockatoos, the black swan, the 'blue' frogs, the dozy koala, the quick skinks and lizards, and sent their manuscripts and drawings for publication in England. Convicts with artistic skills sharpened by the acts of forgery that had sent them to New South Wales were also put to work as natural history illustrators for governors and high-ranking officers to produce the vivid and original wildlife illustrations that resonate with their sense of strangeness today.

But the actual collection and preparation of natural history materials, especially faunal specimens, required a mastery of techniques to which few residents of the infant Colony could aspire. Botanical collection of the unique flora begun so brilliantly by the youthful Banks and encouraged throughout his life, had little counterpart in zoology.

Even so the challenge for botanists was profound. The

renowned British botanist Sir James Smith summed it up. When an observer entered a remote country like Australia, he wrote, he could scarcely meet with any fixed points from which to draw his analogies. 'Whole tribes of plants which first seem familiar prove on a nearer examination, total strangers...and not only the species that present themselves are new, but most of the genera, and even natural orders.'

Robert Brown, the *Investigator*'s talented botanist, destined to become a pre-eminent figure in nineteenth-century science, found this to be all too true as he collected at many points around the Australian coast. The youthful Brown faced problems for which the Linnean system of classification provided no answers. Collecting widely, he gathered a vast diversity of plants through their various stages of maturation and devised his own system of classifying them into families and genera. In doing so, he moved away from the widely accepted Linnean system which was based on the reproductive organs (the stamens and pistils) as the criteria for the basic divisions into classes, orders, families, genera and species, and adopted his own system founded on the anatomy and physiology of the plants' parts.

When Brown's famous *Prodromus Florae Novae Hollandiae* with its 464 genera and 2000 species of Australian plants— all new to science—was published in Britain in 1811, it transformed the study of botany. It offered a reference guide for local residents, and proved the point that classificatory systems fixed with such confidence in Europe were unable to sustain the weight of large new additions from another hemisphere.

The challenge was similar but more demanding in zoology.

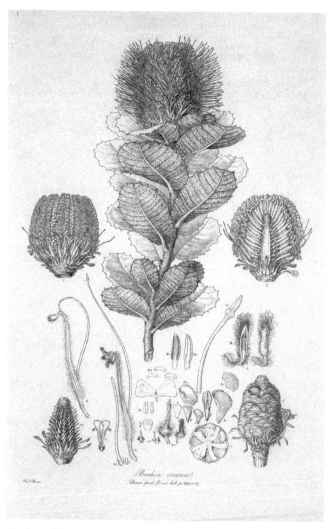

Ferdinand Bauer's black and white drawing of the Banksia coccinea *prepared for, and published as accompanying illustrations to the botanist* Robert Brown's Prodromus of the Flora of New Holland, *1813. The illustration showed the singular nature of Australian flora.*

'Koala or New Holland Sloth' from a watercolour by an early naturalist artist in New South Wales, John Lewin. Australian fauna startled the European gaze, but British comments were far from flattering. George Perry, in his Arcana *published in London in 1811, conveyed the prevailing philosophy: 'As Nature provides nothing in vain', he wrote of the koala, 'we may suppose that even these torpid, senseless creatures are wisely intended to fill up one of the great links of the chain of animated nature, and to show the extensive variety of created beings which God has, in his wisdom, constructed'.*

Few early colonists possessed scientific training—their library resources were small—and few, if any, were acquainted with the theoretical concepts about Australian fauna circulating in scientific circles in Britain and Europe. Hence, for a good two decades of the nineteenth century, anatomists and theorists overseas had to rely on the rough descriptions and very occasional specimens sent them by British travellers and agents visiting New South Wales.

One colonist, however, whose information could be relied on was the high-flying authority and influential pastoralist, Sir John Jamison. A trained medical man, Jamison had served in the Royal Navy and won his knighthood when, as a physician serving on a hospital ship with the Baltic Fleet in the Napoleonic War of 1809, he succeeded in curbing a serious outbreak of scurvy.

In 1816, from his own observations, he communicated to the Linnean Society of London offering specific information about the platypus. 'The female', he conveyed succinctly, 'is oviparous, and lives in burrows in the ground, so that it is seldom seen either on shore or in the water'.

Sometime later, in January 1821, Patrick Hill, a naval surgeon on active service responding to Admiralty directives to naval officers to send back useful observations on natural science, also wrote to the Linnean Society from Sydney reporting on his dissection of a female *Ornithorhynchus* in which he found 'in the left ovarium a round yellow ovum about the size of a small pea' with two of similar size 'and immense number of minute vesicles'. These he sent to Oxford University for examination. Although the egg had not been laid, Hill reported from talk with an Aboriginal elder 'that it is a fact well known to them that the animal lays two eggs about the size, shape, and colour of those of a hen; that the female sits a considerable time on the eggs in a nest which is always found among the reeds on the surface of the water'.

These early fragments of information moved remarkably

close to the truth. Yet a barrier existed between theorists in Britain and the raw information plucked from the Australian field. British naturalists were inclined to indulge a lofty view of the Colonies—of monotremes, marsupials and men. They perceived Australia as 'a faunal backwater', a kind of 'zoological penal colony', and, while they strove to carry off laurels of scientific classification and to build personal reputations on their interpretations of specimens that reached them, they viewed Antipodean aberrations and inversions as a taunt to the hard-won truths of European science.

To the Englishman especially, Australia was a land of deviations and contrarieties. It departed from their norm. There, convicts—the very outcasts of their own society— grew rich as emancipists and became influential in the land. There, signs of real prosperity and social mobility marked a country where the refuse, the dregs of civilised society, were consigned. It was not just the plants and animals that veered so sharply from their European kind.

The visiting French zoologist, François Péron, writing of his experience, had added his measure of doubt and incredulity for a place he felt impelled to call the 'unparalleled continent'. For Péron too, the sense of a reversal of physical and social norms went hand in hand. The civilising effects of trans- portation even in 1802 struck him with awe, while the natural phenomena of this 'singular country' were 'equally strange and incomprehensible'. He contemplated the un- predictability of the natural world and noted that 'amongst the vegetables and animals, nature has multiplied her singularities'.

When the visiting Irishman James O'Hara came to dash

off his impressions of New South Wales in 1817, it seemed
to him also that nature had surely 'indulged in a whim' in
Australia. 'She sometimes mimics herself', he wrote, 'in giving
to smaller animals, such as the native rat, the general form
and characteristics of the kangaroo; she gives to a number of
species the false belly of that animal; in numerous instances,
animals were discovered which might at first sight be
considered monstrous productions, such as an aquatic
quadruped, about the size of a rabbit, with the eyes, colour
and skin of a mole, and the bill and web-feet of a duck, a
parrot with the slender legs of a sea-gull, a skate with a head
like that of a shark'. 'The whole animal creation', he concluded,
'appeared to be different from that of every other region'.

Even sophisticated long-term English settlers responded
sharply to the country's contrariness. When a small elect
coterie of scientifically-minded men met in Sydney in 1821
under the patronage of their astronomer-governor, Sir Thomas
Brisbane, to form an embryonic Philosophical Society of New
South Wales, they deplored the lack of local attention given
to research while they fretted at the capricious and singular
conundrums of the country.

> Upwards of thirty years have now elapsed, since the colony
> of New South Wales was established in one of the most
> interesting parts of the world, interesting as well from the
> novel and endless variety of its animal and vegetable
> production, as from the wide and extending range for
> observation and experiment ... Yet little has been done to
> awaken a spirit of research or excite a thirst for information
> amongst the Colonists. When we consider that we are

speaking in the nineteenth century, and reflect on the progression of science for nearly three thousand years, the rejection and adoption of various systems in every branch of natural history, and the security which it was fancied that scientific arrangement had at least attained, we are almost inclined to believe that Nature has been leading us through a mazy dance of intellectual speculation, only to laugh at us at last in this fifth continent.

The society's moving spirit, Barron Field, an active and cultured lawyer who had been in the Colony for several years, plunged into poetry to sum up his thoughts on a country that defied—even superbly cheated—nature's normal laws.

Kangaroo, Kangaroo!
thou Spirit of Australia,
That redeems from utter failure,
From perfect desolation,
And warrants the creation
Of this fifth part of the Earth,
Which would seem an after-birth,
Not conceiv'd in the Beginning
(For God bless'd His work at first,
And saw that it was good),
But emerg'd at the first sinning,
When the ground was therefore curst;—
And hence his barren wood!

Kangaroo, Kangaroo!
Tho' at first sight we should say,
In thy nature that there may

Contradictions be involv'd,
Yet, like discord well resolv'd,
It is quickly harmonized
Sphinx or mermaid realiz'd,...
Or centaur unfabulous,

Would scarce be more prodigious,...
But, what Nature would compile,
Nature knows to reconcile;...

Bounding o'er the forest's space;
Join'd by some divine mistake,
None but Nature's hand can make—
Nature, in her wisdom play,
On Creation's holiday.

For howsoe-er anomalous,
Thou yet art not incongruous,
Repugnant or preposterous...

When sooty swans are once more rare,
And duck-moles the Museum's care
Be still the glory of this land,
Happiest Work of finest Hand!

Over and above his versifying, Field was a serious contributor to the Colony's science. Australia might be a land of contrarieties where the laws of nature seemed reversed, but zoology was no longer a study of dried birds. 'Australia's zoology', he contended crisply in his writings of 1825, 'can only be studied and unravelled on the spot, and that too only by a profound philosopher'.

Such a message found scant resonance in professional circles at the metropolis. Ships carrying assortments of Australian animals, dead and alive, plied their way to England for distribution to zoological gardens and museums. James Hardy Vaux, transported as a convict to New South Wales and recording his impressions in 1819, wrote that his ship was so crowded with kangaroos, emus and black swans that it resembled a Noah's Ark. Given their problematic place in the systematic scheme of things, museums overseas had difficulties in arranging some Australian items in their displays. As a result, a stuffed platypus appeared in one provincial museum's display cosily naturalised among creatures of British woods and fields.

By the 1820s most major museums in Europe sported a stuffed Australian animal or two procured for them by British agents. Barron Field, in Sydney, frustrated at the lack of zoological expertise or instruction available in New South Wales, decried the way that many private specimens were allowed to be purchased by *foreign* museums. He fed the patriotic zeal. 'It is surely a national disgrace', he declaimed, 'that *we* should conquer and acquire colonies, and that *other nations* should reap the honour of their zoological history'.

Yet the problem remained, information from all sources from the freakish Colony was treated in Britain with prejudice and caution.

During 1828, as the debate on the question of the platypus's mammae simmered in Europe, Peter Cunningham, the Royal Naval surgeon-superintendent of convicts in New South Wales, writing a popular travel book on his two years in the Colony, added further credence to the local view that the platypus laid eggs. It was, he recorded, a 'remarkable animal which

forms the link between the bird and the beast, having a bill
like a duck and paws webbed similar to that of a bird, but
legs and body like those of a quadruped, covered with coarse
hair and a broad tail to steer by'. It was 'believed to lay eggs,
a nest with eggs in it of a peculiar appearance was some time
ago found'. Cunningham, who had lived in the Colony
intermittently over several years and observed the platypus
in its natural habitat, as well as dissecting a specimen, was
the first from the field since Governor Hunter to report that:
'It [the male] bears a claw on the inside of its foot having a
tube therein, through which it emits poisonous fluid into the
wounds which its claw inflicts; as, when assailed, it strikes
its paws together, and fastens upon its enemy like a crab.'

Other visitors crowded in to give testimony. Dr John
Henderson, a surgeon in the Indian Army visiting the Colonies
from Bengal in 1829 added his opinion. Accompanied by his
colourfully-clad Hindustani servant, he sighted the platypus
in the rivers of New South Wales and Tasmania on frequent
occasions, observing that the little creature swam low in the
water, dived rapidly, but was tiresomely difficult to shoot.
Henderson was *au fait* with the debate about the animal's
reproductive processes, whether viviparous or oviparous. But
which was right? All was second-hand. 'A gentleman residing
near the river Hunter [in New South Wales]', he offered, 'has
assured me that he once dissected a female within which
several eggs were discovered of different degrees of maturity'.

Invigorated by Tasmania's cool air and its effects on his
delicate constitution, Henderson flushed out a group of
scientifically-minded residents of Hobart and launched the
Van Diemen's Land Scientific Society for the purpose of fixing

72

their attention on the puzzles of natural science. Launched under vice-regal patronage, the society attracted a visiting British natural scientist, a Fellow of the Royal Society of London, Matthew Friend, who moving between Sydney and Tasmania, carried a fistful of introductions from the Royal, Zoological and Geological societies of London with requests to establish scientific correspondents in the Colonies. Their 'land of mystery' with its many anomalies, he reminded the new members, would, if verified and described, 'tend much to illustrate many of the most abstruse and important questions in the history of organic life'.

Yet few, if any, of the scattered company of amateur naturalists—the resident clergymen, doctors, surveyors, explorers, administrators—from their respective locations in New South Wales, Tasmania and Victoria, were then in a position to investigate the remarkable new species or offer informed perspectives on unknown forms. Cut off by distance, they avoided speculative hypothesis and left sorties into categorisation and theory to scientists abroad. Even verification and specific description were, in the case of zoology, apt to meet difficulties.

The Tasmanians admitted the challenge. When shortly after the demise of the Van Diemen's Land Scientific Society, a more enduring Tasmanian Society of Natural History was formed in Hobart in 1839, they took Blumenbach's *Ornithorhynchus paradoxus* as their emblem and added to the title page of the society's journal the Latin motto *Quocunque aspicias hic paradoxus erit* (From wherever you look at it, this will be a paradox). On the cover of their first minute book an anonymous scribe jotted pithily: 'All things are queer and opposite.'

THE FIRST
HARD LOOK

Those only who are accustomed to view,
and investigate the varying productions of
nature...can appreciate the true feelings
of enjoyment experienced on seeing in their
native haunts creatures which before were
known only from vague description.

George Bennett, 'The Water-Mole
of Australia', 1835

Merchant and naval vessels that plied their way from the
United Kingdom into the far corners of the Indian
and Pacific oceans in the nineteenth century carried in their
ranks of doctors and naturalists a breed of enterprising men
whose minds and eyes were astir for the wonders of the New
World.

George Bennett, born in 1804 amid ships' masts at Plymouth,
was a young man who sought adventure. Educated at a local

grammar school, his inquisitive mind flew out beyond the musty scent of classrooms and in 1819, at the age of fifteen, he sailed before the mast to Ceylon and the island of Mauritius.

Back in London in the 1820s, he studied anatomy and medicine at the Middlesex Hospital and the Hunterian School of Medicine and gained a diploma in medicine at the Royal College of Surgeons where he forged a friendship with a young contemporary, Richard Owen, which was profoundly to influence his career.

In 1829, his medical degree behind him, Bennett was back again at sea. This time he travelled as ship's surgeon to the trading vessel *Sophia* which carried him, in charge of 92 male convicts to Sydney, and on to New Zealand, the Pacific Islands and Asia.

From his days at the Royal College of Surgeons, Bennett was aware of the platypus interest—Owen was dissecting his first 'old' specimen at that time, and on his short stay in Sydney he heard from local sources that 'the platypus laid eggs'. But as he wrote later 'the majority preferred forming theories of their own and arguing on their plausibility, to devoting a few leisure days to the collection of facts by which the question might be set at rest forever'.

It was a hasty judgment as he would later find.

But as an aspiring naturalist, Bennett himself was not idle. When his ship stopped at Erromanga in the New Hebrides (now Vanuatu) to raid its sandalwood forest, he plucked a live specimen of the rare Pearly Nautilus, known previously only by its shell, floating in the waters of the bay, and achieved a scientific coup. He had bagged a living fossil—the sole living genus of the molluscan subclass Nautiloidea (which included the fossil Ammonites)—which was of great interest to systematists.

While Bennett, back in England in 1831–32, presented his specimen to the Hunterian Museum and wrote of his discovery in the *London Medical Gazette*, Richard Owen seized on the scientific potential of the delicate organism and published a superb descriptive memoir on the Pearly Nautilus (*Nautilus pompilius* Linn.) the following year. Modelled on Cuvier's systematic work, it placed the youthful naturalist at the forefront of comparative anatomists and won him coveted election to the Royal Society of London.

Bennett also had much to gain. He was present at the museum when Owen dissected the specimens of the five female *Ornithorhynchus* that led to his conclusions on the mammary glands and watched his friend inject the mercury into the tiny lacteal ducts. The platypus mystery was etched on his mind.

In May 1832 Bennett was on the ship *Brothers* bound for Sydney. This time, he donned his naturalist's suit. Spending only a brief time in the busy Sydney township, he rode out in September across the Great Dividing Range, the first trained observer to search out and report on the elusive platypus. Journeying some distance through the green-grey countryside and finding the vegetation 'melancholy', he got his first sight of the animal at Goodradigbee on the Murrumbidgee River.

'On perceiving it', he recorded:

> the spectator must remain perfectly stationary, as the slightest noise or movement would cause its instant disappearance, so acute is its sight or hearing, or perhaps both; and it seldom reappears when it has been frightened.

By remaining perfectly quiet when the animal is 'up', the spectator is enabled to obtain an excellent view of its movements; it seldom, however, remains longer than one or two minutes, playing and paddling on the surface, soon diving again and reappearing a short distance above or below.

Its mode of swimming is very singular and not always alike; sometimes the body of the animal, beaver-like, is partly raised above the surface, while at others, particularly in the still pools, every part is submerged except the upper surface of the bill and nostrils, and these being but sufficiently elevated above the water to enable the animal to breathe, it is only by the little rings which this operation creates upon the surface that its presence can be detected.

He was also struck with wonder. When seen in a living state running along the ground, he wrote, the animal 'conveys to the spectator an idea of something supernatural, and its uncouth form produces terror in the minds of the timid: even the canine race ... stare at them with erect ears, and the feline race avoid them'.

At twilight on the Murrumbidgee, he shot a female which, wounded, died that night. Next day, he dug out a live female from its burrow. Determined to get this piece of scientific evidence in the best possible condition back to England, or at least to where he might dissect it under good conditions, Bennett rode off with the little animal attached to his mount in a box. A hundred kilometres on, the travelling platypus escaped.

September through October are the platypus breeding

months in the cooler southern regions of Australia. Early in November, Bennett was again in the field, retracing his steps across the Razorback Mountains to the Goulburn Plains and back through the Murrumbidgee regions where his first find was made. And there at Yass, his perseverance paid off.

At a secluded spot on the Yass River, he secured a female with what he suspected to be mammary glands, and at another place, on the 'ponds', three young ones in a burrow. The young died soon after capture. But on his return route to Sydney near Goulburn, Bennett found a young male and female and a grown female and, setting up a rough 'platypus box' crowded with his catch, grass and assorted worms, he started back on horseback.

The platypus bill has been described as the most remarkable organ for sensory perception found in the animal kingdom. Moist, soft and highly flexible (quite unlike the hard bill of a duck) the whole is covered with a dense array of pores leading to highly specialised and sensitive 'touch-corpuscles' which give the animal an acute extra sense. For creatures with such delicate apparatus, the horse's jog-trot and the crowded box proved traumatising and the grown female died. Bennett dissected and preserved her on the spot and then carried the two young platypuses back to Sydney.

At 28, George Bennett now saw himself as a man with a mission. The first European naturalist to capture four highly elusive platypuses from their burrows, he planned the impossible: to tame and domesticate the two young animals and take them as scientific prizes to England. From our perspective, in 2000, the difficulties are recognisably monu-mental. But in 1832, he was a pioneer. Intrigued and hopeful, he kept the two small creatures in a prepared box at his residence in Sydney for almost five weeks, releasing them for periods to scud about his rented rooms.

After their 300-kilometre journey on horseback, he reported in a letter to Owen in February 1833:

> the young animals became much thinner, altho they appeared to feed well on the bread & water and chopped meat, which I gave them. One, the female, died a few days since from the effects of a fall (for they run about my room) the male is living, lively & seems to be regaining flesh. It is a most interesting, harmless animal, but restless & difficult to keep quiet, when once awake; but they sleep much during the day and also night, only coming out of their <u>birth</u> a short time to feed and play about the room.

The diet of the platypus has subsequently been proved to be a huge daily intake of small crustaceans and river insects. Not surprisingly, the young male did not survive.

Bennett preserved the specimens and sent them to Owen. Itemised they included 'the Uteri etc. of the female Ornithorhynchus showing an egg in the left uterus which is laid open'. 'This preparation', Bennett added, 'was left unopened until I arrived in Sydney when I laid it open before Mr. Macleay'.

A second bottle contained the 'organs of generation in the female Ornithorhynchus showing eggs in the left uterus, but smaller in size than in the preceding specimen...the uteri laid open and eggs taken out', also 'the uterus, detached from the surrounding parts, showing eggs (almost in the first stage)'. Still another preparation showed 'the gland and duct communicating with the spur in the Male Ornithorhynchus';

the poisonous spur in which venom is secreted in mature males. Despatching this rich haul of fauna on the ship *Brothers*, Bennett tucked in other specimens of interest to fuel his friend Owen's continuing interest in the generation of the kangaroo. One—a kangaroo uterus showing the foetus with placenta attached—was, Bennett thought, decisive of the animal being brought forth in the usual way', rather than, as was still widely suspected, being born in the 'false womb' of the mother's pouch.

Bennett was right. On this evidence, Owen published a key paper on the generation of the kangaroo in 1834 which announced *his* discovery that the kangaroo young developed in the uterus and not from the nipple in the pouch, thus stamping his name in this zoological field. Bennett, with a good score behind him, zig-zagging back to London that same year, was awarded the gold medal of the Royal College of Surgeons for his zoological presentations.

In 1835, with one more sea voyage behind him, the inveterate traveller put up his shingle as a practising physician in Sydney. At the same time, at the invitation of the Colonial Secretary of New South Wales, Alexander McLeay, a man of science himself, the energetic Bennett took on the secretaryship of the small, recently founded Colonial Museum in Sydney and set about cataloguing its collections.

In 1835 Bennett also published his major descriptive paper 'Notes on the natural history and habits of the *Ornithorhynchus paradoxus* Blum.' in the *Transactions of the Zoological Society of London* which, for the first time, presented a detailed eyewitness account of the living platypus and its ecology to the scientific community.

Bennett's lucid prose described the appearance of the curious beak and head, once thought to be a freakish fraud, on the live animal. 'The head is rather flat,' he relayed, 'and from the mouth project two flat lips or mandibles, resembling the beak of a *Shoveller Duck*, the lower of which is shorter and narrower than the upper, and has its internal edges channelled with numerous *striae*, resembling in some degree those seen in the bill of a *Duck*'. The central portion of the mandibles was a 'bony continuation from the skull, and anteriorly and laterally a cartilaginous substance, perfectly movable, extends from the bony portion to the distance of 5/8ths of an inch'.

Most distinctive was a transverse fold or flap of the membrane/layer at the base of both lower and upper mandibles. Home, from his dissection early in the century, had considered that the platypus used these folds to prevent the beak being pushed beyond this part into the soft mud when foraging. But Bennett, studying the animal's actions on the spot, saw the flaps as affording protection to the eyes as the animal burrowed in the mud. The eyes themselves, set high in the head, were very small and brilliant.

Describing the varying state of the ova found in the uterus of a number of female platypuses he had dissected, he found they were uniformly spherical in form, deep yellow in colour, smooth in their exterior surface with the yellow-coloured yolk occupying different parts of the ovum according to its size. Such first-hand data extended the fragmentary pieces of information that had earlier come to hand: but much to his disappointment, Bennett saw no rudiments of an embryo in any specimen.

Nonetheless his information on the mammary gland seen

in a newly killed specimen was especially timely and confirmed the accuracy of Owen's interpretative dissection. At first, Bennett recorded, 'I was rather surprised to observe scarcely any appearance of it'. In one female specimen who had just produced her young the glands were very large, vascular on the surface and the mammary artery crossed them 'in most beautiful and distinct manner'. But there was no projecting nipple.

Bennett's success in securing platypus specimens—male and female, impregnated, the young in their burrows—and the detail of his fresh observations and dissections at once marked him as an invaluable naturalist in the field and secured his election as a Fellow of the Zoological Society of London.

But how to catch a platypus? Both scientific and popular readers were soon to learn:

The opinion that I had heard advanced at Sidney [sic] of it being requisite to shoot the *Water-Moles* dead instantly, otherwise they would sink and not reappear, I did not find to be correct in practice . . . if the animal is wounded, it immediately sinks, but soon reappears on the surface of the water some distance beyond the place at which it was seen to dive. Some require two or three shots before they are killed or so severely wounded as to enable them to be brought out of the water; and they frequently evade being captured, even when wounded, by frequent and rapid diving. Sometimes, too, unless a sportsman is very vigilant, they may come up among the reeds and rushes, which are plentiful in some parts . . . I have no doubt, also, that some which sink after being wounded, escape into their burrows;

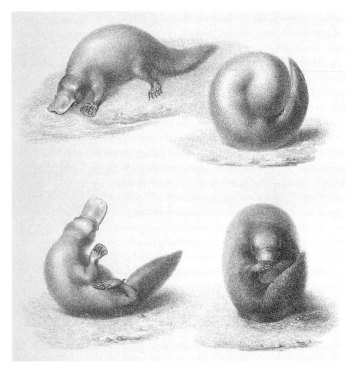

Bennett's first-hand experience brought new knowledge. Here he depicts the platypus in various positions—awake, sleeping, combing itself and eating. The Australian poet Banjo Paterson later described these habits in his 'Old Man Platypus': 'Safe in their burrows below the falls/They live in a world of wonder/Where no one visits and no one calls/They sleep like little brown billiard balls/With their beaks tucked neatly under.'

as even when they cannot reach the bank, they may get access to the hole by the subaqueous entrance.

Central to his early endeavours in locating and securing platypus specimens was the aid provided by the Aborigines.

The enterprising Bennett already had some experience of Indigenous peoples when he began his forays in Australia. In the New Hebrides at Euramonga, he had gallantly rescued a local girl threatened with death by her kinsmen and had taken her back to England.

It was an Aborigine at Yass who introduced him to that vital platypus habitat, the burrow, revealing its entrance concealed among the long grass about a foot from the water's edge. The burrows were long and labyrinthine. In searching them, Bennett observed that the Aborigines with great skill and economy of action did not lay the burrow entirely open, but 'delved holes at certain distances and introduced a stick to ascertain direction previously to again digging down upon it'. This labour-saving method enabled them to explore the whole length of the burrow to its broad oval-shaped nesting chamber at the end, without the angst of opening it entirely. 'The whole of the interior', Bennett noted, 'was smooth, extending about twenty feet in a serpentine direction up the bank. It had one entrance near the water's edge, and another under the water, communicating with the interior by an opening just within the upper entrance.'

These complex burrows, the platypus's shield against a threatening world, were the naturalist's greatest obstacle. Dug a foot or more below the earth's surface, they became narrower as they receded from the entrance which was itself only about the usual size of the animal when uncontracted. The first one he examined extended for a distance of three metres. This in itself was a skilful excavation for an animal a mere 23 to 25 centimetres long but for whom its powerful muscular feet and claws acted as a counterpoint. Other burrows Bennett found

George Bennett's sketch of the burrow of Ornithorhynchus *with the
nesting burrow at the rear.*

later extended 'in a serpentine form' a distance of twelve
metres from the entrance, and one or two more, an astonishing
sixteen metres.

Finding local Aborigines living along the banks of the
Murrumbidgee, and later at Yass, Bennett concluded that
they could shed light on the habits of the animal and especially
its manner of giving birth and caring for its young. He
recorded their names for the platypus, *Mallangong* and *Tambreet*
from tribes in the Murrumbidgee area, names he carefully
added to the varying scientific nomenclature of the platypus
in his catalogue of the collections of the Australian Museum.

But on the subject of the *Tambreet* birth processes, he drew confusing stories.

The Yass Aborigines, he recorded, 'at first asserted that the animals lay eggs, but very shortly afterwards contradicted themselves'. Testing the reliability of their evidence, Bennett drew an oval egg which they agreed was like that of the *Mallangong*, and then a round egg which was also declared to be of the *Mallangong*. The Aborigines said that '"old woman have eggs there in so many days" (the number of which they did not know) that the young ones "tumble down", and that two eggs are laid in one day'. It was difficult going.

Torn between their evidence and that of Aborigines from a different tribe in Tumut country who said that, '"Tambreet no make egg (corbuccor) tumble down; piccaninny make tumble down"', he was none the wiser as to whether the young were born live, hatched from an egg within the platypus ovoviviparously, or if the egg was laid for hatching. The broad testimony of the Aborigines was that the young were found in the burrow 'as if just brought forth'. But, Bennett added, 'from the difficulty they find in expressing themselves in one language, we often misunderstand them'.

Bennett's dilemma in considering Aboriginal testimony mirrored that of earlier enquirers such as Robert Brown. Brown, responding to Banks' hard pressure for information, had also questioned Aboriginal people about the platypus birth process, and informed Banks that from this questioning, he had nothing to add. In fact, he had. A surviving draft of Brown's letter to Banks of 1803 reveals that he omitted a telling paragraph from his final letter which read: 'From the vague reports collected from the natives it might be supposed that the eggs are hatched out of the body.' But Brown admitted: 'I have already learned to be very careful in depending upon what they assert in such cases.'

Patronising British attitudes to the Australian Aborigines went deep. On first contact, Cook had commented on the Aborigines' simplicity and happiness, although he noted at Botany Bay that they appeared 'to want nothing more than for us to be gone'. But his naturalist, young Banks—following his idyllic Tahitian days—judged the first Australians as only one degree removed from 'brutes'. Graphic depictions of the Indigenous inhabitants, idealised by the engraver's art, intermittently conveyed a sense of the 'noble savage' in his 'primitive debasement'. But one thing was sure. Both settlers and visitors came together on the point: Australia's first inhabitants were 'unreliable'.

Bennett concurred, although his interest in Aboriginal health and customs was strong and his respect for the people and adoption of their words for animals and places was more serious than that of many of his contemporaries. Yet, caught between conflicting evidence, he was careful when seeking information about the platypus birth, 'not to ask any person who had been repeatedly questioned before on the same subject'. For, as he noticed, there was a marked tendency among the Aborigines to please their interlocutor. It depended on how, and when, the question was asked.

Yet for Bennett, exact and valuable data sometimes came his way unsolicited, wrapped in 'pidgin speak'. On one such occasion, when dissecting a female platypus for the elusive mammary gland, he recalled in his later writings, that 'a native was overlooking me . . . Perfectly aware, although I had not informed him, for what I sought, he pointed out its situation, saying at the same time, "Milliken (milk) come all same as from cow". When I told him that I could hardly see it, he replied, "When piccaninny come cobbong (large) milliken come"'.

On the question of platypus generation, however, Bennett's mind was pretty well made up from his dissection of the captured platypus he sent to Owen in 1832 in which he had found loose ova in the uterus. Owen figured and described this specimen in a paper to the Royal Society in 1834 and the conclusion took shape in both men's minds that the platypus did not produce its young like other mammals viviparously, by live birth, but that it either laid eggs or was ovoviviparous, producing its young from eggs hatched within the body.

Their preference for the ovoviviparous mode, consolidated over a great span of years, had marked consequences for biological science.

THE PAPER WAR

The need to categorize gave free play to the
imagination of men of science.

Umberto Eco, *Kant and the
Platypus*, 1999

George Bennett's paper vividly illuminated the living
appearance, anatomy and much of the ecology of the
platypus when it was published in Britain in 1835, and for
some three-quarters of a century it remained the primary
descriptive source. Yet even as he laboured in New South Wales,
debate on the 'missing' nipples and the alleged mammary
glands of the perplexing animal reached something like fever
pitch in Europe and England.

Geoffroy, with a sense of infallibility and determination
that escalated in France's distinguished zoologist with age,
expelled a scatter of papers during 1833–34 assaulting Owen's
(and Meckel's) conclusion that the glands of the platypus
were mammary glands and arguing that a young animal with
a mouth like a bird could not possibly suckle from a breast

that lacked a nipple. These, he proclaimed roundly, were *not* mammary glands.

Ticking off his objections in a series of papers which he placed in the *Transactions of the Zoological Society of London*— the journal that Owen himself had founded—Geoffroy argued that the liquid produced from the site was certainly not milk. The most that could be expected from glands of such simple structure as the platypus displayed was, at best, a nutritious mucus eaten by the young. Possibly it was a secretion dispersed in the water for the young after hatching. Or, alternatively, like the musk-glands of the shrew, the liquid was designed to attract a mate. Finally, it was conceivable that these 'Monotrematous glands' as he carefully dubbed them, secreted carbonate of soda to form a shell around the egg. Owen, chided the French anatomist from across the Channel, should make the necessary chemical analysis of the secretion.

Confident for 30 years that the monotremes were oviparous, an egg-laying duo of a reptilian nature belonging to the Monotremata, the distinct order of Vertebrata which he himself had created, Geoffroy threw his hat into the ring. Through all his propositions, of one thing he was sure: oviparity decreed an absence of mammae. It was wholly unnatural to have a mammiferous, milk-producing, egg-laying animal—an amphibian mammal. Worse, it went against the 'all-powerful rules', the very plan of Nature. 'I merely wish to play a useful part', he added to his own defence beguilingly in his 1833 paper on 'New Observations on the Nature of the Abdominal Glands of *Ornithorhynchus*', 'restricting myself to the obligation of a naturalist having the privilege of age, confident in the

The French zoologist, Etienne Geoffroy St-Hilaire.
A spirited and controversial participant in the platypus
debate. 'If these are mammary glands', he demanded,
'where is the butter?'.

experience of older studies, and acquainted with the possible
variations in nature in order to assist observers less practised
than myself in the study of natural history'.

Richard Owen was up to the feint. With a number
of consummate papers behind him, he already walked
authoritatively on the biological stage. He was a brilliant
zoologist and, quipped his contemporaries, he had 'brains
enough to fill two hats'. The Royal College of Surgeons and
its Hunterian Museum were the key centre of comparative
anatomy in Britain. He had married the daughter of the
curator, William Clift, and his own position as assistant

curator gave him a dominance of the museum's collections. He also had access to other museums and to a diverse spread of animals that were given to dying at the London Zoo.

Owen, too, was already beginning to act as a magnet for information from collectors around the world. His friend Bennett's eyewitness evidence from Australia was a key buttress to his factual store. The 'milk gland' was very large, Bennett had confirmed in his letter of February 1833. 'I can now inform you from actual observation that milk is secreted from it: it comes out, (as your mercury did when you injected the ducts) in small ducts on the surface of the skin.'

There was other reinforcement. During 1834 Owen was able to examine the 'neonates'—the two baby platypuses— which Lieutenant Maule had sent him from New South Wales. With these under his knife, he established two definitive facts: that the suckling infant's mouth was not duck-like in its embryonic form but designed to allow it to imbibe nourishment from the glands in normal style; and secondly, that the substance found in the babies' stomachs was milk.

Blainville, who had succeeded to Cuvier's Chair of Comparative Anatomy at the Museum of Natural History in Paris following the master's death in 1832, rallied behind the mammary glands. A comparative anatomist of high distinction who looked for order in the chaos of nature based on principles anchored firmly within God's design, he argued that, regardless of the substance the glands secreted, they must be identified as mammary glands since they performed the same nourishing function.

Round one had gone to Owen. Yet in the paper war with Geoffroy, Owen looked for bridges. Although the elderly and

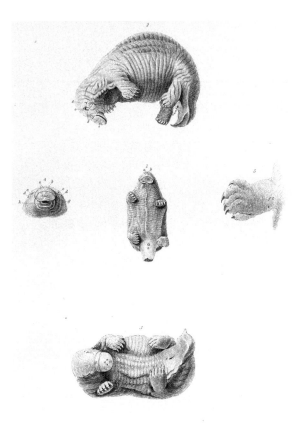

The platypus neonate. Lithograph drawing by Rymer Jones illustrating Richard Owen's paper 'On the young of the Ornithorhynchus paradoxus, Blum'.

overbearing Frenchman managed soon afterwards to block Owen's election as a Foreign Correspondent of the prestigious Institut de France, Owen offered a partial olive branch.

His own anatomical dissections, his mercury injecting experiment, and his correlation of the size of the mammary glands with the changes in the ovaries of female platypuses, set down in his detailed papers and his responses to Geoffroy, all pointed to the place the monotremes occupied as mammals within the classified order of nature. His findings on the echidna supported him. But he was willing to stress the affinity between the Mammalia and Geoffroy's Monotremata which emerged clearly in their circulatory, respiratory and nervous systems.

In this Anglo-French paper war, Owen emerged victorious. His careful empirical research prevailed. The monotremes were accorded a clear place within the definition of mammals. These highly aberrant Australian animals could now be given a firm taxonomic position within the class Mammalia, pushing out its boundaries and demonstrating for all systematists how wide the variation of natural types within a class could be.

Yet much was still to be done. Despite new anatomical knowledge, the monotremes's method of generation remained the crucial unknown: oviparous as the elder Home, Geoffroy and Lamarck decreed; viviparous as many other naturalists believed; or ovoviviparous—incubating and hatching their young within the mother's body. Only precise information on the egg, provided by naturalists in the field, Owen repeated, could determine the creatures's process of reproduction.

Yet, even as he exhorted, his own preference was clear. From his various dissections of the platypus, Owen favoured ovoviviparous reproduction and contended that the failure to find eggs was itself indicative of their dissolution in the mother's body. He judged moreover that it was not now necessary to believe that a mammiferous animal must necessarily be viviparous. A belief in the animal's oviparity, as he saw it, had also only been necessary *before* the confirmation of the mammary glands when it could be assumed that the egg yolk was required to provide nourishment for the young.

Underpinning the differing scientific opinions held by naturalists was a sweep of philosophical and theological ideas that coloured responses to organic classification and to changing interpretations of nature's origins and laws. The concept of the 'great Chain of Being' which organised organic life in a ladder of fixed development from the lowest and humblest forms up to man had reached its zenith in the first half of the eighteenth century and its language could still be heard among some early nineteenth-century savants.

But by the first years of the nineteenth century, Cuvier's innovative thrust into comparative anatomy had dealt it a heavy blow. From his influential position at the Paris Museum of Natural History, he introduced the concept of four major divisions of organic life—Radiata (jellyfish and starfish), Articulata (worms and insects), Molluscs (snails, octopus) and Vertebrata—to account for a complexity that could not be accommodated in a single plan that progressed from the simplest to the most complex. A master of comparative forms, he saw organic life as an immense network in which species depended on each other. Abandoning the Chain of Being, Cuvier introduced the concept of the correlation of parts, based on his recognition that organisms were created with their organs so shaped as to enable them to adapt to their conditions of existence.

It was a static world and, in his view, no longer one where

a progressive tendency through adapted forms was seen as part of a Divine purpose.

Researching in 1810 among the strata of the Paris Basin, Cuvier brilliantly applied the principle of the correlation of parts to the fossil record. There, to the astonishment and delight of peers and public, he found that the smallest fragment of bone, however insignificant, possessed a fixed and determined character that related it across the board to the class, order, family, genus and species of the animal to which it belonged. Thus by careful analogy and comparison, working from a fossil tooth here, a jawbone or femur there, it became possible to reconstruct a striking array of gigantic and extinct animal forms.

The new discipline of palaeontology ushered in far-reaching change. Large new questions confronted biologists and geologists. How in a universe of design could these extraordinary resurrected creatures have been formed? Had they all existed at the beginning of life on earth, altering in size and form due to a changing physical environment over long eras of time? Or had they been created by a divine hand at specific times?

Cuvier's answer lay in catastrophism and separate creation. While as an orthodox Christian, he had originally believed that his 'immense network' had been fixed from the six days of Biblical creation, from his fossil researches he now perceived that reptiles had appeared on land long before mammals, that now-extinct species were the first in the earth's record, and that it was only in the most recent geological strata that fauna appeared almost identical to that of the present day.

Cuvier, the 'Mammoth', large in mind and substance, now seized on the view, as several geologists had before him, that physical catastrophes culminating in the Biblical Flood had repeatedly destroyed organic life which was replaced after each catastrophe by successive creations. With each separate creation, new and more complex, but unrelated, species appeared.

Lamarck saw the situation differently. An expert in invertebrates, a professor with Cuvier at the Paris museum, he conceived the natural order as classes arranged in a linear fashion from the most complex descending to the least developed forms. Nature, he argued, having constructed the simplest animals and plants directly, produced others from them in spontaneous generation. Changing physical needs and environments led to new responses which sparked new habits in the organism. Modifications acquired in an organism's lifetime were then passed on through reproduction and the inheritance of acquired characteristics.

When he confronted the puzzle of the fossil record, Lamarck rejected extinction but underlined the importance of fossils in revealing past change. Species, his theory of transformism ran, were transmuted into their living descendants through change effected over very long periods of time. Through this lens, Lamarck rejected catastrophism and argued for the continuity of past and present. His followers called it Lamarckian evolution.

These divergent positions strained and sundered the international community of science. Cuvier deplored his colleague's theory as an affront to Christian belief. Geoffroy, also working at the museum, accepted the transmutation of species but not Lamarckian transformism. Cuvier's former pupil Blainville, no longer a disciple or admirer of his master, held to the Chain of Being while allowing for changes in species and classes. From his own palaeontological researches, Blainville proclaimed that gaps in the chain of existing organisms would be supplied by fossil forms. 'Together', he postulated, 'extinct and extant organisms testified to the fullness and rightness of God's creation'. In this chain of completeness and rightness, Blainville assigned the Australian platypus to the base of the descending order of organisms from man, through the placentals and marsupials, to the monotremes.

Across the Channel in Britain, a slew of viewpoints flourished. The distinguished Professor of Geology at Oxford, William Buckland, who (unlike many geologists of the period) did not consider it possible to reconcile geology with the Book of Genesis, believed that new forms of organic life came directly from the intervention of the Creator. He also wrote with passion and erudition about the catastrophic Universal Deluge and its impact on the history of the earth.

By 1830, the youthful geologist Charles Lyell gripped attention and drew converts to his new theory of uniformitarianism, a theory challenging to catastrophism and founded on conclusions from his own geological observations that changes in the earth's crust occurred slowly and cumulatively by forces still in action over vast periods of geological time. He launched his persuasive theory in his *Principles of Geology* published that year. Lyell, however, rejected organic progressionism and found Lamarck's evidence 'fanciful'. Yet, while he recognised that species could become extinct as a result of failure in the struggle for existence and saw that extinct species were replaced by others, he fell back before any evolutionist explanation.

The ideas of Archdeacon William Paley, spelled out in his *Natural Theology* of 1802, had a compelling hold on men's minds. Paley's view of the universe implied an intelligent and benign Creator whose work was everywhere to be seen. A bird's feathers for flight, webbed feet for travel in water, pouches for carrying young across long and dry distances, the stretching tongue of the woodpecker for catching grubs were all instances of design. Design implied a designer. 'That person', Paley proclaimed, 'is God'. In his lucid prose, the confident theologian drew on the analogy of the watchmaker, who creates an intricate machine and sets it in motion, to illustrate his idea of complete and regulated design. His book was prescribed reading in the universities.

In this intellectual climate where theology was intricately linked with geology and natural history, men turned to miracles to explain any new and unexpected phenomena they encountered. Creationism, catastrophism and separate creation where a purposeful God sacrificed whole series of his creatures only to reinvent them afresh at different times in other centres of development, were concepts that were reappraised only slowly as the century advanced.

In this environment, Richard Owen became one of the pre-eminent scientists of Britain. Deeply influenced by Cuvier, whom he visited in Paris in 1830 two years before the 'Mammoth's' death, he honed his skill in comparative anatomy and four years later applied his mind to fossil remains when invited to describe the fossil bones which Darwin brought back from South America after the *Beagle* voyage. Propelled by a workaholic's zeal, Owen's researches stretched out across the British Empire and into the rich fossil reservoir of the earth. In a century of unfolding ideas and new and far-reaching classification, he stood as the impresario of the fossil record, 'the British Cuvier', reconstructing a vast cavalcade of huge extinct fauna from teeth and bones dredged from British, Australian, New Zealand, African and South American soils.

Launching his early zoological investigations profoundly influenced by Cuvier and fiercely opposed to transformism,

Richard Owen, c. 1846: 'a tall man with great glittering eyes', as his friend, the historian Thomas Carlyle, described him. He holds a fossil bone of one of the giant extinct creatures he was brilliantly recreating.

Owen emerged increasingly in his palaeontological researches as a successive and continuous creationist. He held that each species had been created only once in time and space, but that its diffusion was the result of its own law of reproduction under the enlarging or restrictive influences of external circumstance.

Nature, he acknowledged, had advanced with 'slow and stately steps' offering a unity of plan in the plant and animal kingdom. This plan offered both clear relationships between species and, in the case of vertebrates, furnished a distinct 'archetype' of which all vertebrate animals were variations on a theme. Webbing his argument together from a maze of living and extinct forms, Owen came to see these variations as expressing a distinct advance. Working assiduously, he marked a purposive and repetitive structure in the animal kingdom, which, he contended with certainty, flowed from a 'beneficent Sovereign of the Universe, an all-wise and powerful First Cause'.

Richard Owen represented the cutting edge of British establishment science. But tucked within his huge research output, God's purposive plan for the Australian monotremes had yet to unfold.

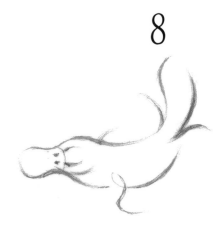

DARWIN'S
PLATYPUS

A Disbeliever in everything beyond his own
reason, might exclaim, 'Surely two distinct
Creators must have been at work'.

Charles Darwin, Diary, 1836

O n a cold day on 7 December 1831, Charles Darwin, aged
22, a newly minted graduate of Cambridge University,
set off as unpaid naturalist to Her Majesty's Ship *Beagle* on
her four-year voyage of navigation and survey and to establish
chronometrical stations around the world.

Young Charles was not a particularly obvious choice for
the protracted task. A timely serendipity intervened. Educated
at Shrewsbury School where he considered his education
'simply a blank', Darwin was a late developer. He attended
Edinburgh University Medical School, swooned at operations,
hated medicine, independently picked up some natural history
and the principles of classification and, contemplating a life

Charles Darwin in 1840, watercolour by George
Richmond, Down House. For the young naturalist, the
diversity and geographical distribution of species he observed
in the southern hemisphere during the voyage of the Beagle
transformed his thinking on the natural world.

in Holy Orders, transferred to Cambridge. There he battled
with mathematics, became interested in entomology, was
charmed and convinced by the writings of Archdeacon William
Paley and his argument about a universe of God's beneficent
and orderly design, and studied geology. There too, he gave
an assenting tick to the prevailing geological concept of
catastrophism.

Darwin joined the *Beagle* through the say-so of his professor at Cambridge, John Henslow, who, when offered the prospect of a long sea journey of scientific discovery, declined the job himself. Young Darwin, known as the grandson of the famous Dr Erasmus Darwin and judged 'the best qualified person . . . likely to undertake such a situation', was made the offer not on the grounds, as the professor explained, of Darwin's being 'a *finished* naturalist' but one 'amply qualified for collecting, observing, and noting anything worthy to be noted in Natural History'.

So worked the old and new boy's network. And such was the discerning insight of Cambridge's Professor of Botany.

It was thus that Charles Darwin, the most promising naturalist of the age, came to view the amazing and improbable platypus in its habitat and to enlist it to his cause.

After several years at sea with landfalls in South America and the Galapagos Islands and brief stops in Tahiti and New Zealand, Charles Darwin arrived in Sydney in January 1836 on the homeward leg of the *Beagle*'s survey, hungry for mail. With no letters from home to cheer him, the long voyage seemed to him wrapped in tedium. 'The time spent in making passages [the chronometrical surveys]', he wrote to his sister disconsolately, 'is to me so much existence obliterated from the page of life. There never was a ship so full of homesick heroes as the *Beagle*.'

Stopping in Sydney only long enough to notice that this was 'a wonderful colony; ancient Rome, in her Imperial grandeur, would not have been ashamed of such an offspring', Darwin, adopting his usual practice, hired a horse and man to guide him and rode off across the Blue Mountains. Crossing the Great Dividing Range by the road convicts had carved through the high mountain spine of New South Wales, he stayed overnight at a large pastoral property on the Cox's River—'Wallerawang'—to whose manager he carried a letter of introduction.

'In the dusk of the evening', he wrote in his Diary, 'I took a stroll along a chain of pools (which in this country represents the course of a river) & had the good fortune to see several of the famous Platypus or Ornithorhynchus paradoxicus [sic]. They were diving & playing in the water; but very little of their bodies were visible, so that they only appeared like so many water Rats. Mr Browne shot one; certainly it is a most extraordinary animal. The mounted Specimens', Darwin concluded, observing the soft and pliable bill of the live animal, 'do not convey a proper idea of the head and beak; the latter being contracted and hardened.'

The experience clearly moved Darwin. Writing a few days later to his next host, Admiral Phillip King, one of the Colony's most pre-eminent scientists whose son was a midshipman on the *Beagle*, he recounted, 'In the evening, we went with a gun in pursuit of the Platypi & actually killed one . . . I consider it a great feat, to be in at the death of so wonderful an animal'.

Darwin viewed the fabled creature with a mind that was not entirely unprepared. He was aware of the classificatory struggle. He had read his grandfather's *Zoonomia* which had briefly, and rather wildly, pondered the mating habits of Australia's curious beasts. In Edinburgh in the 1820s, he had made the acquaintance of the prominent radical naturalist Robert Grant who was himself scientifically interested in the platypus and who, on walks with his young student, talked to him about Lamarck's theory of evolution. It was a theory Darwin met again, although this time under severe attack, in Charles Lyell's *Principles of Geology* which he carried on the *Beagle*. Volume two—to his delight—reached him at Rio de Janeiro. Reading these two fat scientific best-sellers on the voyage, Darwin absorbed Lyell's theory of the infinitely long, gradually changing development of the physical world set, as the author saw it, against a universe of static organic creation. In the dry Antipodean landscape, the young naturalist came to ruminate on the star player in Australia's faunal drama. 'Earlier in the evening', he jotted in his Diary at Wallerawang station, 'I had been lying on a sunny bank & reflecting on the strange character of the Animals of the country as compared to the rest of the World. A Disbeliever in everything beyond his own reason, might exclaim, "Surely two distinct Creators must have been at work"'.

But, moments later, Darwin noticed the conical pitfall of an ant-lion and observed how it shot jets of sand at its prey exactly as the same genus, though a different species, did in the Old World. Now he asked, 'what would the Disbeliever say to this? Would any two workmen ever hit on so beautiful, so simple & yet so artificial a contrivance? It cannot be thought so', (he answered himself). 'The one hand has surely worked over the whole world.' Paley, it seemed to him then, was right. It was the perfect design of 'one overseeing Designer', although he was ready to note, 'A Geologist perhaps would suggest, that the periods of Creation have been distinct & remote the one from the other; that the Creator rested in his labour'.

Several years of research and thinking lay ahead before Darwin's careful yet thrusting mind fully marshalled the

Plate 1   Ferdinand Bauer's drawing of two platypuses was the first scientifically accurate depiction of the animal to be made from observation in the field, although it was not brought to public view through publication until late in the twentieth century.

bear on the last joints of the hinder legs, the under parts of which are firm and ca... lous like a hoof, their tail which is strong & muscular they use as a weapon of defence, and when pursued very close they annoy and hurt the dogs very much with it.

...is the most curious and singular animal I have ever seen, the one from which I have taken the drawing is one foot, eleven inches & half from the point of the bile th... to end of the tail, eleven inches in the round of the thickest part of the body and fou... ...ches & half in the length of the tail from the anus, it is web footed, on the fore foo... ...four nails or claws and ... behind like the gaff of a ... not placed immediatl... ...at extremity, on its hind foot is five claws remarkably sharp and long and

Plate 2   *The platypus sketched informally in his journal by John Washington Price, surgeon to the convict transport ship* Minerva *on its voyage from Cork, Ireland, to Sydney, 1798–1800. Price conveyed some deeply held European views of the animal kingdom when he scribbled in his Journal: 'If we examine through the various regions of the earth, we shall find that all the most active, sprightly and useful quadrupeds have been gathered around man, and either served his pleasure or still maintained their independence by their vigilance, their cunning or their industry. It is in the remote solitudes that we are to look for the helpless, deformed and monstrous works of nature...'*

Plate 3   *Charles-Alexandre Lesueur's delicate drawing of the echidna made during the Baudin expedition to Australia in 1802–03. Geoffroy St-Hilaire placed the echidna with the platypus in a new and separate taxonomic class, 'Monotremata', in 1803.*

Plate 4   *Ships on their way back to Britain from New South Wales often called in at Calcutta, where specimens of dried and preserved fauna were known to be presented to the great collector, Marquess Wellesley, Governor-General of India from 1797–1805. A later successor, Lord Moira, Marquess Hastings, was also an eclectic natural history collector who, on a journey from Calcutta to the Punjab during his Governor-Generalship from 1814–18, assembled drawings of Indian animals made by the 'Bengali draughtsman' who accompanied him, Sita Ram. This pastel drawing of the platypus by Sita Ram, preserved in the Hastings Album at the British Library with no note of its origin, was either drawn from a dried specimen or, less likely, modelled on depictions of the animal available in the British literature. The platypus had penetrated the Orient.*

Plate 5   *Convicts were among the first to record some early impressions of the
curious 'wonders of nature' in Australia. T. Richard Browne, a former forger,
incarcerated at a secondary penal settlement at Newcastle, New South Wales,
conveyed a vivid sense of the oddity and strangeness of the animals he painted for
the Commandant between 1812–21.*

Plate 6   *Platypus country, Tasmania*

Plate 7   *John Gould's plump platypuses, sketched during his*
excursions collecting mammals and birds in eastern Australia in 1838–40
and published in his Mammals of Australia, *were designed both for science*
and the market place.

Plate 8   *A platypus photographed underwater. Such photographs are difficult to obtain in the wild given the animal's passion for privacy.*

evidence on isolation and modification of species as major agents of evolution and organic change. Yet his Australian experience fed the process. In a remote property in New South Wales, he had sighted Australian 'crows like our jackdaws', a bird like the magpie—birds as similar to English birds of the same name that plainly belonged to different species— a miniature kangaroo in size and locomotion like an English rabbit, and the 'wonderful animal' with an immediate resemblance to the water-rat, seen on a summer evening. Here were examples of animals similarly adapted but widely separated geographically which belonged to different orders, families, genera and species.

The platypus remained etched in his consciousness. Observing the grotesque African plant of the desert *Welwitschia mirabilis* with its long coiled leaves writhing like a vegetable octopus later at Kew, he dubbed this floral mixture of the primitive and advanced, 'the vegetable *Ornithorhynchus*', the platypus of the plant kingdom. The advanced yet primitive mammal at the Antipodes, the sleek and small elusive platypus, drifted in Darwin's maturing ideas on biogeography and diversity.

Back in England, the young man who had gone forth wrapped in the prevailing theological and creationist philosophies of his period returned a scientist committed to empirical evidence and his own inquisitive, synthesising ideas. Getting his absorbing account of *The Voyage of the Beagle* quickly into print, from 1837 Darwin began to jot down his seminal 'Transmutation Notebooks' based on findings from the voyage shored up by his voracious reading.

Yet even before he began to write his famous Notebooks,

he had, from his own eyewitness observations, already begun to question Paley's static, purposively designed universe and the immutability of species, and to shape the concept of transmutation. Victorian naturalists' wide acceptance of separate creations and their belief that a new and successive creation followed each major and extinguishing upheaval of the earth, appeared to him riddled with holes.

Piling up a stack of weighty tomes around him, Darwin plunged into the writings of his contemporaries—in botany, zoology, palaeontology, geology, plant and animal breeding—laboriously seeking facts. In 1837, as his evolutionary theory took shape, rereading his friend Lyell's *Principles of Geology* which saw species as fully determined and fixed, he scribbled jauntily, 'if this were true adios theory'. He was equally resistant to Lamarck's notions of evolutionary adaptation and progressionism spelled out in his *Philosophie Zoologique* (1830), scrawling in his own copy, 'Very poor and useless Book'.

After four years exploring diverse landscapes around the world, Darwin was more alert than any of his predecessors and contemporaries to the relevance of the geographical distribution of species and the effects of isolation on their modification and range.

Working on his great 'Species Book' for nearly twenty years before committing it to print, he drew the platypus into his synthesising scene. Writing to the young botanist Joseph Hooker in March 1844, he puzzled over those genera which were 'not typical', a feature of the platypus with its peculiar mammae, and concluded that such genera were rendered atypical by the extinction of allied genera.

Darwin was wrestling with the question of 'fewness of

species' and 'aberrant genera' and his need to explain this against the argument of creationists. For creationists, such oddities could be described as products of a creative whim. For Darwin, the *Ornithorhynchus* and echidna with their lack of radiating sub-families were both few and aberrant and they posed problems for the theory of evolution by natural selection. Yet they would surely be anomalous, he reasoned with the doubting Hooker, if they had a dozen species instead of one. Hooker was his sounding board.

Such anomalous forms as the platypus (along with air-breathing Ganoid fishes) Darwin wrote him, unravelling his own ideas, were exemplars of genera that had arrived through natural selection at a lower stage of perfection in their smaller areas of habitation than species inhabiting a much larger area and confronting greater competition. He allied them with species which inhabited fresh water.

Yet writing to his friend Lyell in these early reflective years, he also saw the phenomenon of the platypus as obtaining its beak by 'retrograde recurrent development'. For, as he put it, '[the platypus] united in one organism disparate characters from different embranchments, the bill of a bird, skeletal resemblances to reptiles and hair and body of a mammal'. However, as he pointed out to Lyell, 'Natural Selection acts exclusively by preserving successive slight *useful* modifications, hence natural selection cannot possibly make a useless or rudimentary organ... A nascent organ, though little developed... must be useful in every stage of development... The mammary glands in Ornithorhynchus may perhaps be considered as nascent compared with the udders of cows.'

Darwin's ideas, aired informally in his letters to close

colleagues, would appear coherently in *The Origin of Species* in 1859. Writing lucidly there on the fundamental question of classification, he accepted the correlation of characters derived from parts as the true formula for determining an animal's taxonomic position. By such correlation of a part, even if trifling, naturalists could identify a character which was nearly uniform and common to a great number of forms but not common to others, and use it as one of value.

In such taxonomic determination, the platypus offered an illustrative case. 'If the Ornithorhynchus had been covered with feathers instead of hair', Darwin gave it as a key example, 'this external and trifling character would have been considered by naturalists as important an aid in determining the degree of affinity of this strange creature to birds and reptiles, as an approach in structure in any one internal organ'.

Whatever its confusing taxonomic aspect, the platypus was an explicit player in Darwin's ideas on isolation and species diversity. 'In Australian mammals', he later developed his thoughts succinctly in his eloquent and persuasive book, 'we see the process of diversification in an early and incomplete stage of development'. He noted again the enduring character of organic forms in fresh water. In such environment, new forms, (he went on to suggest) will have been more slowly produced and old forms more slowly exterminated. '[I]n fresh water', he wrote, 'we find some of the most anomalous forms now known in the world, as the Ornithorhynchus and Lepidosiren which, like fossils, connect to a certain extent orders now widely separated in the natural scale. These anomalous forms', he summed up, 'may almost be called living fossils; they have endured to the present day, from having inhabited a confined area, and from having thus been exposed to less severe competition'.

Darwin's platypus, that 'wonderful animal' he had seen shot in a tranquil Australian river, became indeed a strong case for his argument of the survival of the fittest. Retrograde recurrent development it might well exhibit, but it was a staunch survivor. Though Nature granted vast periods of time for the work of natural selection, Darwin reminded readers of *The Origin*, 'she does not grant an indefinite period ... if any one species does not become modified and improved in a corresponding degree with its competitors, it will soon be exterminated'.

Such were the finished products, in 1859, of Darwin's thinking on the monotreme. Yet twenty years earlier Australia also contributed to a major flashpoint of understanding for his broad theory of evolution that broke upon him from the evidence of the fossil record after his return to England in 1836.

The great South American fossil bones he had carried back to London on the *Beagle* proved ignition points. Unearthing these relics of ancient geological times, Darwin had expected that the bones would prove to be European and African rhinos and mastodons, and not specifically South American types. Richard Owen's descriptions and classifications of them in his Appendix to Darwin's *Voyage of the Beagle* brought him up short. As if through a luminous prism, he saw the close relationship between the extinct giant species and the living sloths and armadillos. 'These facts', he noted graphically in his Journal, 'origin of all my views'.

Skirmishing in the literature, he hit upon vital corroborating evidence from the Wellington Caves of New South Wales.

While the *Beagle* had been making passages around the world, the Australian cave discovery of the early 1830s fell like an arrow into tight-held creationist theories in Britain. During 1830, an Australian colonist, George Ranken, fossicking in the distant pastoral country of south-west New South Wales, had discovered the Wellington Caves where, to his astonishment, he unearthed a rich haul of fossils of giant extinct marsupials. Ranken packed some off to the Colony's Surveyor-General, the scientifically-minded Sir Thomas Mitchell. Mitchell eagerly explored the site himself and dispatched a selection of the fossil fragments to the Professor of Natural History at Edinburgh University. These remarkable fossils soon reached the Hunterian Museum in London for the expert examination of William Clift.

Clift's identification of the fossil teeth, jaws and femurs as belonging to giant extinct kangaroos, wombats and dasyurids (carnivorous marsupials) that roamed the Australian countryside in the Tertiary period, at once raised doubts about 'the Universal Deluge'—so ardently advocated by Oxford's Professor of Geology, William Buckland, and other geologists as the major agent of the earth's change—and threw up further question marks from another corner of the world about the concept of geological catastrophism and separate creation.

In the first place, the cave findings furnished initial evidence of a past geographic distribution of mammals in Australia, a distribution that matched evidence coming to light on extinct gigantic mammal forms in Europe. Second, the evidence indicated for all to heed that the genera to which the Wellington Caves fossils belonged continued, with their highly distinct organisation, to inhabit the Australian continent in present times.

The question exposed a real puzzle. For if, as the catastrophists and separate creationists claimed, the animals and plants of the Tertiary era were destroyed by the Biblical Flood and replaced subsequently by a specially created issue of new fauna and flora, why should a clear continuity prevail between existing species and extinct forms?

Surveyor-General Mitchell was alive to his coup. 'I understand Buckland's nose is put completely out of joint by the bones from Australia', he wrote jubilantly to Ranken. 'I have now heard that the fact of the fossil bones belonging to animals similar to those now existing has worked a great change in all their learned speculating on such subjects at home.'

Darwin, for his part, had read an allusion to Clift's findings in Lyell's second volume of his *Principles of Geology* (1833) during the *Beagle* voyage. But it was his access to the detailed evidence in 1837 that firmed his view of transmutation. He would describe it in *The Origin*. 'Mr Clift many years ago showed that the fossil mammals from the Australian caves were closely allied to the living marsupials of that continent.' The South American revelations also showed that 'the fossil mammals buried there in such numbers are related to South American types'. 'I was so much impressed by these facts', he wrote, 'that I strongly insisted in 1839 and 1845 on this law of succession of types and on the wonderful relationship in the same continent between the dead and the living'.

'On the theory of descent with modification', he summed up in his clarifying prose,

> the great law of the long enduring, but not immutable, succession of the same types within the same areas, is at

once explained; for the inhabitants of each quarter of the world will obviously tend to leave in that quarter, during the next succeeding period of time, closely allied though in some degree modified descendants. If the inhabitants of one continent formerly differed greatly from those of another continent, so will their modified descendants still differ in nearly the same manner and degree.

The great leading facts of palaeontology seemed simply, as he concluded, 'to follow on the theory of descent with modification through natural selection'.

For Darwin, the killing of a platypus in New South Wales in 1836 had far-reaching outcomes. It was like a pebble cast into a quiet pool. Its ripples spread. When amid outcry Darwin published *The Descent of Man* in 1874, the platypus would rise again as a key exemplar of natural selection and as a diversified link in what he perceived as the organic chain of mammals rising up to man.

While the reproductive processes of the platypus and the echidna were still unknown in 1874, Darwin saw the Monotremata as 'eminently interesting, as leading in several important points of structure towards the class of reptiles' and as structural precursors of the marsupials, placentals and on to man. For the world, as the great evolutionist reasoned, appeared as if it had long been preparing for the advent of man. 'He owed his birth', said Darwin, 'to a long line of progenitors...If any single link in this chain never existed, man would not have been exactly what he now is.'

9

TO THE
ANTIPODES

Australia is the habitat of this curious creature
[*Ornithorhynchus anatinus*] which, though
plentiful, is still very little known...No
further progress has been made towards the
solution of the still pending and highly
interesting physiological question—Does the
Platypus lay eggs?

> Gerard Krefft, Curator of the
> Australian Museum, *Australian
> Vertebrates; Fossil and Recent*, 1871

W hile Darwin moved the complex pieces of his biological
mosaic into place across the 1840s and 1850s, Richard
Owen swung expansively to his creative reconstruction of
extinct birds and mammals from fragmented bones being
dug up by collectors from strata around the world. At the
same time he maintained what had become his proprietorial
interest in Australia's living fossil. For the 'great Richard', as
the Australian Museum curator rather impatiently called him,

the solution of platypus generation loomed as a key classificatory question and as a central query in his unfolding of nature's plan.

As acknowledged expert, Owen had summed up existing biological knowledge of the *Ornithorhynchus* in his essay on 'Monotremata' in the *Cyclopaedia of Anatomy and Physiology* in 1841. In his concise style, he noted that the monotremes 'present the extreme modifications of the Implacental type, and mark the last step in the transition from the Mammalian to the Oviparous classes'. His skilled observing eye pinpointed the physical functions and appearance of the animal, opened (largely from his own dissections) its cranium, tongue, stomach, beak, glands, urogenital tract and all other parts to view, and itemised its singularities. The animal glowed from the learned page. 'The body is clothed with a dense coat of hair', Owen wrote, 'which consists of a fine fur, intermixed with long, stiff, flattened and sharp pointed hairs that seem to represent the spines of the Echidna'.

> The feet are short, broad, armed each with a fine claw, but less robust than an echidna. The four feet have a web, which not only unites and fills the interstices of the toes, but extends beyond the extremities of the claws; the web of the hind foot terminates in the base of the claws. With these swimming-feet is associated a strong, broad, horizontally flattened tail, which completes the organic locomotive machinery for the aquatic existence of an airbreathing and warm-blooded quadruped.

Here was the master's word portrait of the animal he judged to be a vital connecting link between mammal and

bird. There were gaps in knowledge of some of its functions—
the horny pointed spur on the heel of the male (a feature shared
with the monotrematous echidna) was such a one. Anomalous,
like other features of the animal, its suspected use as an
'offensive weapon' (which Bennett for one questioned) had yet
to be proved. Above all, the mode of generation and course
of development of the Monotremata, 'although elucidated in
many essential points by the light of anatomy and analogy',
he observed, required 'observation of the breeding animals,
and of the impregnated uterus and embryo in several stages,
before they can be fully determined'.

On the metropolitan cusp of the northern hemisphere far
from the animal's preserve, it was a point Owen was prone
to repeat.

Yet little enough was happening in the Colonies. Bennett,
taken up with his rising medical practice and work at the
Colonial Museum, was no longer in the field. Owen must stir
other pots. In the early 1840s he turned to the itinerant
Surveyor-General, Thomas Mitchell—known to him from the
Wellington Caves fossils—and flattered him with an invitation
to assist, in the course of his inland surveys, 'to solve some
of the most puzzling enigmas in Physiology'.

It was a full assignment. A female *Ornithorhynchus*, the distant
scientist specified, should be shot and pickled every week during
the months of September, October and November: '100 would
not be too many' he urged, 'if we could thereby settle the long
disputed question, oviparous, ovo-viviparous, or viviparous?'.

'A hole should be cut in the belly', his instructions ran,
'and spirit thrown in, and then she should be placed in spirits
bodily. If the keg is two-thirds or more filled by specimens,

*Dr George Bennett in later years. The Australian-
based zoologist involved himself in a field search for
evidence of the platypus's reproductive process for more
than half a century.*

the spirit should be changed once or twice before they are
finally packed off—otherwise it becomes too weak to preserve
the internal parts necessary to be dissected for the purposes
of the enquiry.'

But events hastened slowly in the Colonies. Mitchell's
next major survey in 1845 took him away from platypus
country to the dry regions of northern Queensland. A decade
later, confined by administrative work in Sydney, the Surveyor-
General was dead.

It fell instead to ornithologist and zoologist John Gould, on an eighteen-month visit to Australia in 1838–40, to inspect and report on the animal that still held the London scientific world in thrall.

Gould, a skilful organiser, something of a British Audubon, was out to blend science with the market-place. Trained as a taxidermist, he had already achieved notice as a writer on *Birds from the Himalayan Mountains* and *Birds of Europe* before he ventured to the Antipodes. Short, cheerful and immaculately dressed in black suit and top hat, with his gun and collecting bags slung across his arm, he travelled from Tasmania to South Australia and into New South Wales and employed other collectors to extend his wide collecting net.

Working rapidly, he gathered a mass of information on the habits, economy and appearance of the Colonies' birds and mammals. The birds he gave to his gentle illustrator wife, Elizabeth, to draw and paint with elegant accuracy while he made rough sketches of mammals and preserved their skins for precise depiction and engraving by a team of artists in the publishing enterprise he set up on his return to London.

In his *Introduction to the Mammals of Australia*, Gould described his first close encounters with the monotremes. 'Tired by a long and laborious day's walk under a burning sun', he wrote, 'I frequently encamped for the night by the side of a river, a natural pond, or a water-hole, and before retiring to rest not infrequently stretched my weary body on the river bank; while thus reposing the surface of the water was often disturbed by the little concentric circle formed by the *Ornithorhynchus*, or perhaps an echidna came trotting towards me'.

John Gould, splendidly attired, on a collecting foray in the Australian bush during his eighteen-month visit to the Colonies, 1838–40. Oil painting by Henry Williams, c. 1839.

He was more formal in his account in *The Mammals of Australia* which he published in thirteen parts studded with beautiful colour illustrations in the years 1845–63. There, reflecting on Australia's 'paradoxical creations' in 1863, he noted that 'unquestionably the most singular and anomalous of all these animals was the Ornithorhynchus, with the habits and economy of which, as well as the mode of its reproduction, we are even now, after an interval of fifty-five years [since its discovery] but imperfectly acquainted'.

Gould had digested Owen's writings on platypus anatomy and physiology and soaked up Bennett's eyewitness account. Nonetheless he was well aware that much more remained to be made known about 'this extraordinary type among quadrupeds'. From his own observations and specimens received from other places, Gould recorded its presence in Tasmania and in eastern Australia from Port Phillip to Moreton Bay in the streams and rivers flowing on both sides of the mountain ranges, and in the tributaries of the Murray and Darling rivers.

But his news was gloomy on sites nearer the settlement of Sydney. In the Hawkesbury and Hunter rivers there was, he recorded, a diminution in platypus numbers 'due to the wholesale destruction dealt out to it by the settlers, which, if not restrained, will ere long lead to the extirpation of this harmless and inoffensive animal'. It was, Gould said, often killed 'from mere wantonness, or at most for no more useful purpose than to make slippers of its skin'.

Gould's observations were sharp and well informed. Writing for the general reader rather than the scientific naturalist, his aim was to show that each great division of the globe had its own peculiar forms of animal life and that the fauna of Australia was widely different from that of every other part of the world. 'To account for this on specific principles', he admitted, writing before Darwin's thoughts were published, 'would be very difficult', though he observed that the law of representation was widely apparent among her mammals.

While the practical Gould avoided theory, he inclined to the view—drawing on his knowledge of the marsupials—that the platypus was ovoviviparous. 'On my returning from Australia', he reported in *The Mammals of Australia*, 'the venerable Geoffroy St-Hilaire put the following question to me, "Does the platypus lay eggs?", and when I answered in the negative, that fine old gentleman and eminent naturalist appeared somewhat disconcerted'.

'Now, this oviparous notion', Gould went on, 'was nearly in accordance with the true state of things—somewhat akin to what is actually the case; and I consider the most striking peculiarity of this singular animal, and indeed of all the Marsupiata, to be the imperfectly formed state in which young are born'.

> The Kangaroo at its birth is not larger than a baby's little finger, and not very unlike it in shape: in this extremely helpless state, the mother, by some means at present unknown, places the vermiform object to one of her nipples within her pouch or marsupium; by some equally unknown process, the little creature becomes attached by its imperfectly formed mouth to the nipple, and there remains dangling for days, and even weeks, during which it gradually assumes the likeness and structure of its parents; at length it drops from this lacteal attachment into the pouches, reattaches itself as hunger prompts it so to do, and as often tumbles off when its wants have been supplied.

Ovoviviparity in the platypus, Gould assumed, would produce young which were recognised as being little more than embryonic in form. 'This', he concluded roundly, echoing what theorists like Owen and others had asserted years before, 'is a very low form of animal life, indeed the lowest among the Mammalia, and exhibits the first stage beyond the development of the bird'.

Gould's research on the Australian mammals, with its beautifully depicted, now famous illustrations and his detailed notes on each mammal, issued in parts across nearly twenty years, marked the first attempt to bring these unique fauna collectively under public review. While there was some murmuring among London's scientific elite, as Gould marketed his lavish productions, that he 'was not quite a scientific man', his superb books sold and carried an eyewitness account and an elaborate depiction of the popular platypus to a wider world.

While Gould charmed British readers with his representations of the platypus and canvassed the ovoviviparous birth, in Australia at the turn of the half-century, an anonymous poem, 'The Land of Contrarieties', publicly proclaimed a different generative mode:

There is a land in distant seas
Full of all contrarieties.
There beasts have mallards' bills and legs,
Have spurs like cocks, like hens lay eggs.

But where was the proof? No definitive evidence of an egg travelled back to anatomists and theorists in Britain and

Europe. Moreover, by 1850, the original key protagonists in the platypus mystery were dead: Lamarck in 1829, Everard Home and Cuvier in 1832; Meckel in 1833; the controversial Geoffroy St-Hilaire in 1844, Blainville in 1850. The younger Owen, now the undisputed prince of comparative anatomists, inherited the systematists' unanswered question.

Owen, however, was baulked by scientific silence in the Colonies. Despite his appeals and overtures, return on the monotremes was scant. Members of the lively Tasmanian Society of Natural History furnished a nil return. Their *Tasmanian Journal of Natural Science*, under its apt platypus crest and motto, shone in the years 1842–46 as the one scientific journal in the Australian Colonies, but Owen's published plea to members inviting them to round up and pickle impregnated female specimens of the animal for despatch to him in London was the only reference to the platypus in the journal.

The self-declared Polish count and naturalist, Paul de Strzelecki, traversing New South Wales' and Victoria's eastern flank, threw his hat into the ring. Noting in his account of the region's zoology in 1848 that 'striking and wonderful peculiarities of external and internal organisation distinguish them', he observed that in the case of the duck-billed *Ornithorhynchus*, covered as it was with fur, moving on four webbed feet, suckling its young, it was 'most probably viviparous' and had been discovered to possess a series of contrivances by which it was fitted to live equally well in the elements proper to two distinct classes of animals. This creature, 'a world of wonders in itself', said Strzelecki with a respectful nod to the pre-eminent savant, 'has been found by

Professor Owen to approximate to the reptiles in its generative system'. Both viviparity and ovoviviparity remained on the cards.

British opinion in this scientific matter held the lead; but the French remained alert. During the 1840s, M. Jules Verreaux, one of the special brand of 'Naturalistes Voyageurs' which the National Museum of Zoology in Paris trained and sent forth to different parts of the world to collect animals for France's scientific museums, arrived in Australia with the platypus particularly in mind.

Verreaux spent several years in the Colonies and for some fifteen months studied the habits of the platypus in Tasmania. He found them abundant in the rivers of New Norfolk and killed several in the high altitude of Mount Wellington near Hobart. French to his bootstraps, he was the first to record the manner of platypus coitus:

> During the month of September I came to discover that the coupling took place in the water. Carefully hidden under an expressly made canopy, and in the depths of which it was necessary for me to stay entire nights without daring to move, for the *Ornithorhynchus* is of an excessively distrustful nature, I was able to follow all its movements.
>
> The male, having followed the female for more than an hour, always finished by leading her into the middle of the reeds. There, clinging firmly with the help of his beak, he held strongly on to the skin of her neck while applying the spurs to the posterior part. The female, while struggling energetically, would swim and utter plaintive squeals which bore some resemblance to those of a little

pig and which were ever increasing: the coupling lasted
five or six minutes and then the two animals would play
together for more than an hour. [Author's translation]

Verreaux also cast refreshing light on the early development
of the embryonic young and gave the first record of young
platypuses suckling in the wild. 'At the end of fifteen to
twenty days', he relayed, the once-naked young 'are covered
with silky hair, and are able to swim'. Their beak, he said,
'did not at all recall the form of that of the adult, but was
short, broad and thick, and could embrace in that state the
mammary areola concealed by the hairs of the mother'. When
they suckled, his acute eye discerned, they 'continually triturate
the mother's belly with the fore-feet'.

Watching them play with their mothers at some distance
from the bank, he distinguished 'that when they wished to
procure [nourishment] they profited by the moment the
mother was among the aquatic plants, and where there was
no current. The female having her back exposed, on the
exercise of a strong pressure, the milk would float to a little
distance, and the young might suck it up with facility; this
it does turning about so as to lose as little as possible.' 'I have
witnessed this fact repeatedly, both daily and nightly', Verreaux
wrote. 'I have also remarked that the young, when fatigued,
climbed upon the back of the mother, who brought it to the
land, where it caressed her.'

Throughout his Tasmanian sojourn, the French naturalist
had, more than any previous reporters, observed a considerable
number of young and adult platypuses in their natural habitat.
Returning to Paris in 1848 with an assorted collection of

Australian fauna, he published his observations in the *Revue Zoologique*.

Not surprisingly, Owen seized on Verreaux's paper and translated it in part for a paper, 'Remarks on the "Observations sur l' Ornithorhynque" par M. Jules Verreaux' which he published that year. Wet but observant among the reeds and rivers, Verreaux had proved an astute field-observer; but, to the English scientist's disappointment, he had failed to produce a foetus *in utero* as the ultimate proof of the animal's generation.

While Owen, from his own long dissecting experience, clearly favoured the platypus's ovoviviparity, the Frenchman's evidence was very welcome. 'The number of *Ornithorhynchi* which I have possessed', Verreaux wrote, 'has perfectly demonstrated to me that this animal does not lay eggs, as has been supposed, but that it is ovo-viviparous. The ovaria, which form part of my collections, sufficiently prove this.'

Science, however, waited on decisive answers. The wait was already long. Yet for the proprietorial Owen, his only recourse was to repeat his request to the Tasmanian naturalists for impregnated uteri from females killed at different periods and at different stages of development, asking them to send 'the hinder half of each specimen' to London. He would wait in vain for such 'hinder halves' from the island that sat, like the full-point of a question mark, below the Australian continent.

There was a small flurry of action in New South Wales. In 1850, the scientific P. P. King entrusted several platypus specimens to the care of young Thomas Henry Huxley who, at the outset of his brilliant biological career, had been stationed in Sydney Harbour as surgeon and naturalist of HMS *Rattlesnake* and was about to sail for home following the ship's four-year survey in Australian waters.

From 1855–57, Owen also received two consignments of three female platypuses from a Melbourne resident, J. S. Dismorr, although obtaining impregnated specimens, the obliging colonist confided, was 'particularly difficult'. In 1858 came Dismorr's tired disclaimer that he could procure no more.

But in 1864 Owen was no longer stalled by silence. In that year, a Victorian physician forwarded him an account that a recently captured platypus had laid two eggs. Owen reacted dismissively—much as he had done 30 years before over the account of the eggshell in the debris of the platypus nest described to him by Lieutenant Maule. Then he had been gratified by the suggestion that monotremes gave milk, as Maule advanced in his report of their flowing mammary glands; but he chose to dismiss the reported 'eggshell' as excrement coated in urine salts. The trauma of capture, Owen had said on this occasion, may have caused the female platypus to abort spontaneously.

Sixty years into the platypus mystery, Owen was caught in a paradigm. It was a paradigm largely of his own making. With no other researchers challenging his opinion, the ovoviviparous generation of the *Ornithorhynchus* was judged to be an 'accepted truth'. For Owen, the animal's ovoviviparous mode of birth would establish the *Ornithorhynchus* in its 'true' taxonomic position between reptile and mammal. Essentially, he looked to the Colonies to give his judgment the final tap of truth.

In the event, a key battle in the conceptual war on species and their place in nature would be fought not in the scientific collecting grounds of Australia, but in the realm of philosophical debate in Britain.

10

THE CLASH
OF TITANS

At every significant step along the way, [in the
solving of the puzzle], the obvious signposts were
misread, and misread from maps which
themselves were already out of date.

Jacob Gruber, 'Does the Platypus
Lay Eggs?'

By the 1850s Owen's reputation was second to none in the
biological capital. His industry was enormous. He had
published widely on monotremes, marsupials, Darwin's South
American fossils, the New Zealand moa, a large reference
book on teeth, works on the nature of limbs and the anatomy
of fishes, and he was deeply involved in his studies of fossils
garnered from Imperial outposts around the world.

While his long tapering hands reached deeply into the
arcane fields of the fossil world, Owen's evening lectures
which he delivered each year as the Professor at the Hunterian

Museum, blazed across the grand sweep of natural history and were attended by large and eager audiences.

In 1841, this buoyant scientist had coined the term 'dinosaur', conceiving the great reptile as next to the mammals in life's scale—the highest of the reptiles—and equipped it with a perfect circulatory system and a four-chambered heart. On the crest of Victorian admiration for his awesome fossil reconstructions, he transformed his palaeontological findings into a series of spectacular cement models of giant extinct mammals which were erected in the Crystal Palace Gardens for the London Exhibition of 1854. Before the fascinated gaze of press and public, Owen and his 21 guests sat down to dinner in the model of the Iguanodon.

In 1856, an authoritative and Olympian figure, he took up his appointment as head of the Natural History Department of the British Museum.

In a world where new philosophical ideas were shaping, Owen was the voice of Anglican science. He hobnobbed with bishops and politicians, was the darling of dukes and aristocrats; Prince Albert cultivated him and invited him to Windsor Castle to lecture the royal children in natural history, and Queen Victoria granted him a royal house in Richmond Park where he lived out his long and active life.

Owen's philosophical concepts were firmly drawn. He had early espoused his view of the 'archetype' in vertebrate history, the 'primal pattern' on which all vertebrates were based and which the cavalcade of fossils emerging from the earth's strata confirmed. From his ranging work on fossil forms, he judged that the extinction of species went hand in hand with progressive creation which had continued to furnish a succession of species in which the archetypes were preserved. By contrast, Lamarck's evolutionary 'progressionism' was an abomination to him.

Owen might jettison fixity of species; but he stoutly rejected transmutation. In sum, he believed that each species had been created but once in time and space, but that there was a secondary natural law at work allowing for an adaptive progression of species. On the mechanisms of this adaptation he had noticeably less to say; but he attributed it to the evidence of 'Creative foresight' and to an 'ordained continuous becoming'.

In his lofty position of power and authority, men turned to him for his judgment as challenging new ideas of evolution swirled about the London scene. His opinion, with its strong moral overtones, carried weight. When the controversial semi-scientific and philosophical work *The Vestiges of Creation*, published anonymously by Robert Chambers, stirred the public as seven rapid-fire editions went to print in 1844, Owen praised it privately to its author for its definition of a higher 'generative law' in explaining changes in species, but hurled his strength in public against any argument that linked organic advancement to man.

Owen allotted man a sub-class of his own. To the pre-eminent anatomist who enjoyed the privilege of dissecting deceased apes and chimpanzees from the London Zoo, man was no descendent of their kind. 'He was as different from the chimpanzee', Owen proclaimed in 1857, 'as the ape was from the platypus'. 'I wonder', mused Darwin in the privacy of his Sussex home, 'what a chimpanzee would say to this!'.

Darwin, working in his own time without a public position,

was constructing a very different view. His concept of common descent by modification of species, the existence of single centres of specific origin—with Australian marsupials and monotremes a case in point—and subsequent migration and colonisation, was a far cry from Owen's prescribed world with its successive creations spurred by a generative but divinely initiated 'Second Cause'.

Initially, when Owen first met and classified Darwin's South American fossils, the two men's friendship grew. Darwin's *Voyage of the Beagle*, Owen wrote the young traveller on its publication, was 'as full of good original wholesome food as an egg'. But a growing arrogance and coldness on Owen's part weakened the ties. The unassertive Darwin ('a child', as one observer put it, 'could talk to that man') was drawn instead to a band of younger scientists appearing on the London scene. Joseph Hooker and Thomas Huxley, influenced by their adventurous voyages of survey and exploration in the southern hemisphere, offered congenial company.

For Hooker, Darwin was a touchstone. Even before setting off early in 1840 as assistant surgeon and botanist to HMS *Erebus* on the British Antarctic expedition to Australian and Antarctic waters, he had slept with the proofs of Darwin's *Voyage of the Beagle* (lent to his father by Lyell) under his pillow. 'And no more instructive and inspiring work', he wrote later, 'occupied the bookshelf of my narrow quarters throughout the voyage'. During his time in Tasmania collecting plants for his famous *Florae Tasmaniae*, Joseph Hooker grasped the exceptional nature and distribution of the Australian flora and started on the slow change from his initial belief in

The botanist Joseph Hooker, from a portrait by George Richmond, 1855. Hooker would become Darwin's staunch supporter. Yet, botanising in Tasmania in 1842, he had been ready to entertain the view that Australia's highly distinctive flowers and plants arose, as he put it, 'from a separate creative effort from that which contemporaneously peopled the rest of the globe with its existing vegetation'. His correspondence with Darwin had large effects. By the time of his own book's publication in 1860, Hooker, much influenced by Darwin, had found the explanation of difference and distribution in the descent and modification of species.

separate creation and the fixity of species to a growing acceptance of Darwin's thesis.

It was thus to Hooker, returned from distant places, that Darwin sent the first hint of his theory of natural selection in 1844 and it was the young botanist, rising in the scientific fraternity and his first convert, who overruled Darwin's desire to concede priority for the theory of natural selection to Alfred Wallace and arranged for the double communication of the two men's work to the Linnean Society.

Published less than a year after *The Origin of Species*, Hooker's *Florae Tasmaniae*, which Darwin had discussed as it emerged from his friend's pen, was, said Darwin, 'the greatest buttress to the theory of evolution'.

Thomas Huxley had also put his first foot on the scientific ladder in Australia. Between dancing at balls in Sydney and falling in love with Henrietta Heathorn whom he later made his wife, he garnered the ordinary marine life of Sydney Harbour (the Portuguese man-of-war) to make his great study of the Medusae and to devise an approach of classifying to a basic structural plan that emerged from the study of many individuals. The study, published as a major memoir in comparative morphology in London, won him instant recognition in the scientific community at home and election as a Fellow of the Royal Society at the age of 26.

Working on his papers in London, Huxley became aware that 'the archetypal common plan was a badge of common descent' and that individualism was expressed not in static (as Owen argued) but in dynamic terms. 'How extremely stupid', exclaimed the excited young naturalist when *The Origin of Species* with its mechanisms of natural selection saw print, 'not to have thought of that!'. Articulate, individual, a science communicator of rare skill and zeal, he would become known as 'Darwin's bulldog'.

These two were Darwin's champions. But Huxley, twenty years Owen's junior, was also shaping to give voice to a

T. H. Huxley, surgeon and naturalist to HMS Rattlesnake
*surveying in Australian waters in 1846–50, was destined to
become one of Darwin's key defenders and a leading figure
in British science.*

mounting groundswell of opinion against the powerful
anatomist.

'Owen is both feared and hated', he wrote in a letter in
1851—a year after his return to England—to W. S. Macleay,
the distinguished British systematist who was now living in
New South Wales. 'It is astonishing with what intense

feelings...Owen is regarded by the majority of his contemporaries.' Looking down from a great height, proud and egotistical, Owen for all his authority and prestige was fast becoming the most unpopular scientist in the kingdom. He was also moving out on a scientific limb. Working brilliantly from bone to bone, he was technically immensely knowledgeable, but, as the sharp young Huxley saw it, 'in abstract reasoning, he becomes lost'.

Unlike Darwin, Owen's knowledge of Australian species failed to fertilise his concept of that 'question of questions', species change. In an address to the British Association in the early 1840s, he noted that Australia's fossil record had begun to yield evidence of correspondence between its latest extinct and present Mammalian fauna, a fact, as he said, 'the more interesting on account of the very peculiar organisation of most of the native quadrupeds of that division of the globe'. However, he excluded this evidence of the variation of species and their concentration in particular geographical areas from any overall consideration of adaptive organic advance.

Although ready to adopt the notion of the 'succession of types', a phrase he would later claim (to Darwin's irritation) to be his own, Owen utterly repudiated evolutionary thought. Yet for all his erudition, he could offer no mechanism for the progression of species. He clung instead to environmental factors as the key to species change. And when, in 1859, he advanced the thesis that there was some special factor of the Australian environment that made it uniquely suited to marsupials, Darwin, who knew of similar conditions for marsupials in Brazil, saw it as Owen's 'gigantic hallucination'.

With the publication of *The Origin of Species*, Darwin found

Owen his most bitter foe. Then Darwin's offering was no longer seen as 'wholesome food': he had become the 'Devil's disciple'. In a series of readily identifiable anonymous reviews, Owen fiercely rejected the mechanism of natural selection in the evolution of organic species, referred with scorn to Darwin's 'ingenious suppositions', ignored his scrupulous building of the factual data, and castigated the 'theorising' of the work. 'One of my chief enemies', Darwin wrote sadly in his Diary with rare censure, '(the sole one who has annoyed me)'.

Significantly, the platypus and echidna, so pressing and singular in Owen's physiological and morphological research, failed to penetrate his analysis of Nature's laws.

By contrast, Darwin had explained a great string of previously inexplicable facts and swung his lens on species around the globe. His first edition of *The Origin of Species* sold out within a week, and a second edition was rushed out to waiting readers in January 1860. Yet, even within days of its first publication, Darwin's interpretative thoughts had turned again to the monotremes.

He returned to them in private correspondence. More than any other, Darwin wished to bring the anti-progressionist and creationist Charles Lyell, the man he dubbed 'my Lord High Chancellor in Natural Science', into his evolutionary camp. Their letters, questioning and friendly, sped through the mail.

Lyell inferred progression in Darwin's theory of evolution. Not so, said Darwin. Natural selection did not imply a necessary tendency to progression. 'According to the progressive theory', he responded, 'why sh. there be a living platypus or Ornithorhynchus?'. There were no ancestral platypus fossils of an elementary or synthetic kind and, 'if they should be found, how have they escaped being altered, improved and specialised in 30 periods?'.

Working through *The Origin*, Lyell pressed his old argument for the concept of separate creations, or at least 'a primeval creative power' that did not act with uniformity, to explain links between distinct genera. Darwin was blunt. From homologies, he replied to his doubting friend in September 1860:

> I should look at it as certain that all mammals had descended from some single progenitor. What its nature was, it is impossible to speculate. More like, probably, the Ornithorhynchus or Echidna than any known form; as these animals combine reptilian characters (and in a lesser degree bird characters) with mammalian. We might give to a bird the habits of a mammal, but inheritance would retain almost for eternity some of the bird-like structure, and prevent a new creature ranking as a true mammal.

A few days later, as Lyell wrestled with his case against natural selection, Darwin hammered his point: 'I quite agree with you on the strange and inexplicable fact of Ornithorhynchus having been preserved...I always repeat to myself that we hardly know why any one single species is rare or common in the best-known countries.' Yet, he mused in his letter to Lyell on 23 September 1860, that, while he knew of 'no known animal of a grade of organization intermediate between mammals, fish, reptiles etc. whence a new mammal could be developed', this 'unknown form' was 'more probably related to Ornithorhynchus than to any other known form'.

Darwin's spirited commentary brought results. As he had

long hoped, Hooker, Huxley and eventually Lyell came to his cause, and Darwin felt that, while all the world might rail, his case was safe. The theory of natural selection would prevail.

The clever Huxley would have the last word on the irony of science. The ablest of men, he reflected, was a product of his time, 'profiting by one set of its influences, limited by another'. Such was the brilliant and powerful Owen. His creative scientific memoirs opened up new worlds: his inflexible Anglicanism and establishment attitudes fossilised his philosophical and conceptual views. Outliving his rival, Owen fought publicly to avert Darwin's interment—the burial of the 'Devil's chaplain'—in Westminster Abbey in 1882.

Without success. Tourists of every creed and kind now troop to stand beside Charles Darwin's polished brass memorial wedged in the Abbey floor while his books continue to fertilise biological and social thinking into the twenty-first century. By contrast, the once courted and publicly honoured Sir Richard Owen had forfeited much of his scientific influence by the time of his death in 1892.

SOLVING
THE MYSTERY

The Origin of Species, with its widening circle of acceptance, marked a major shift in people's view of nature, descent and adaptation in the organic world. It would in time become the cornerstone of modern biological science. Yet, entrenched in people's minds, old ideas and allegiances lingered. While the concept of evolution and the infinitely creative force of natural selection pervaded scientific thinking, the puzzle of the platypus had not gone away.

While Owen himself was increasingly challenged by the young Turks of British biology in the debate on species, his dominance of platypus research and theory remained unchanged. Dissecting and lecturing in London, this 'master savant' waited for more impregnated female specimens from Australia. But even while he waited for the final data from the field, he entertained little doubt about the animal's mode of generation.

Several factors convinced Owen of the rightness of his choice of ovoviviparity. The structure of the platypus's generative organs; his own familiarity, from his dissections, with the nature of the ovum; his judgment that the platypus's configuration was not substantial enough to extrude a large bird-like egg, and his knowledge that this curious animal was a mammal (however lowly on the mammalian scale) which gave milk to suckle its young—all predisposed the great anatomist to the premise that the platypus hatched its young from eggs within its body.

Such were the inferences, deduced from his metropolitan dissecting table, that anchored Owen's scientific opinion. He awaited only the impregnated uterus and embryo of the platypus in several stages of development from colonial naturalists and collectors to confirm these anatomical findings.

True, there was a notable want of speed. The relationship between leading metropolitan scientists and their colonial colleagues was interdependent, but unequal. Colonial naturalists of all persuasions amassed data, but, for a great part of the nineteenth century, they left the formulation of theory arising from it to scientists at the metropole.

The first Vice-President of the Royal Society of New South

Wales, the clergyman-geologist Reverend W. B. Clarke, a considerable savant in his own right, put their situation in a nutshell in his inaugural address in 1867. He then advised members of the new society to 'discern clearly, understand fully and report faithfully,' but to 'abjure hasty theories and unsupported conjectures... leaving time for judgment whether for or against us, by those who come after us'.

In this environment, the reputation of so commanding and honoured a scientist as Owen glowed like a beacon in Australia. His writings on marsupials and monotremes touched a small coterie of colonial naturalists. But his brilliant reconstructions of Australia's extinct giant marsupials, birds and mammals dating from the finds in the Wellington Caves of the early 1830s, sent an eager troupe of explorers, surveyors, geologists and museum curators scurrying to dig up fossil fragments for the great man's palaeontological zoo. Their contributions filled out the gathering assembly of huge animal ancestors which had roamed the earth in distant periods of geological time. Owen's melodious prose spurred the investigators on. For were not Australia's more recent geological deposits, he suggested, the very 'grave of many creatures that have not been dreamed of in our philosophy'?

As time went on Owen's old bone collectors numbered most of Australia's leading naturalists. Drawing on fossil relics unearthed from pastoral wells as far afield as Queensland, southern Victoria, Tasmania and western New South Wales, this ready-handed army packed off the materials for the research papers which Owen penned regularly on extinct Australian species and which he published collectively as a major work, *Researches on the Fossil Remains of the Extinct Mammals of Australia*, in 1877.

This formidable accomplishment wrote Owen firmly into the history of Australian palaeontology. He was its pioneer. But his philosophical influence on his contemporary Australian naturalists established a more powerful allegiance.

Owen's concepts of continuous creation, 'successive continuous becoming' as he called it, and of 'the irrefragable evidence of Creative foresight' were readily accepted in Australia. The scientific communities scattered in New South Wales, Victoria, Tasmania and South Australia were made up largely of Oxford and Cambridge men who gave part of their private time to botanical, geological or zoological enquiries and drew to their interests a sprinkling of self-educated settlers. But, distant from the thrust of metropolitan debate, they were slow to accept concepts of evolution.

Many colonial naturalists were great admirers of Darwin of the *Beagle*. Victoria's German government botanist, Ferdinand von Mueller, the most honoured scientist in that colony, confessed to cutting his scientific teeth on the *Beagle* journal. Others like W. B. Clarke gladly corresponded with Darwin after *The Origin of Species* reached Sydney, sending him geological data and conducting botanical experiments to assist the British naturalist in his fact-gathering research. But, almost to a man, the colonial scientists rejected Darwin's theory of evolution by natural selection.

Their letters announced their deep concern. As the respected systematist W. S. Macleay, flushed from reading *The Origin*, exclaimed to his friend Viscount Sherbrooke in England: 'The question is no less "What am I?", a creature being under the direct government of his Creator, or only an accidental sprout of some primordial type that was the common progenitor of both animals and vegetables.' 'It is far easier', he concluded, 'for me to believe in the direct and constant government of the Creation of God, than that He should have created the world and then left it to manage itself'.

The influential von Mueller, no longer so admiring, wrote direct to Owen. He declared his position that—after 22 years of observing Australia's plant life, collecting and classifying botanical specimens from across the continent—he for one was committed to the concept of the fixity of species. He was certain that 'we are surrounded by species clearly defined in nature, all perfect in their organisation, all destined to fulfil by unalterable laws those designs for which the power of our creating God called them into existence'.

Clarke, too, a scientific leader responsible for scores of reports on the geology of New South Wales and a collector of Australian animal fossil bones for Owen, publicly urged an open-minded investigation of Darwin's impressive facts; but he himself held fast to separate creation. Hence, for at least two decades after the arrival in Australia of Darwin's 'dangerous book', the climate of scientific opinion remained strongly resistant to Darwinian thought.

Only that other scientific German, the eccentric and independent-minded Gerard Krefft, curator of the Australian Museum, contesting some of Owen's palaeontological conclusions, corresponding with Darwin and advertising at £50 a time for platypus eggs, became an early Australian advocate of evolution. In the vanguard of Darwinian ideas, Krefft indeed saw in Owen's overarching research on extinct and existing marsupial forms a clear illustration of evolutionary mechanisms in Australia.

Owen's close associate, George Bennett, also subscribed to Owen's philosophical orthodoxy. Bennett had become a man of scientific substance in the Colony. His books, *Wanderings in New South Wales, Batavia, Pedir Coast* and his later *Gatherings of a Naturalist in Australasia* had a large currency. These, along with a swag of journal articles on natural history, established him as an informed writer on animals, birds, marine life and the botany of the Asian and Pacific regions. As scientific institutions in Sydney grew, he became a trustee of the Australian Museum, a member of the board advising the new University of Sydney on science education and the founder and active promoter of the Acclimatisation Society of New South Wales. He had a flourishing medical practice and wrote eclectically on health and medical matters, on tetanus and smallpox, and, in later life, became an examiner in medicine and therapeutics at the University of Sydney.

But, although well recognised as a fellow of the Linnean and Zoological societies of London, Bennett's subservience to his old friend had grown across the years. Tugging a colonial forelock in a style not untypical of naturalists at the Antipodes, he saw himself increasingly as Owen's servitor in Australia. Across his long intermittent years of fieldwork, his assault on the platypus had been considerable. Despite the acknowledged difficulties of securing the obscurely living animal (his letters and those of others hammered the point), his own exertions had notched up a telling score. Yet, while he shot and embalmed a significant haul of female specimens, some which contained loose eggs in the genital apparatus, across several decades for his English mentor, he had not been able to bridge the gap between these findings and the young he found in the burrows.

All that Bennett knew for certain, over and above his charming and informative descriptions of the habits and character of the 'duck-bill' sent off to the Zoological Society of London, and his vain attempts to keep several animals in captivity, was that the infant platypus, conceived somewhere between August and October, emerged in December after an indeterminate period in the burrow, covered with hair, to swim quietly in the river with its mother.

What took him so long to bring any solution to the puzzle?

In his geographical isolation, George Bennett was caught in a conceptual fix. While in early investigations he had initially thought that the platypus might prove to be either viviparous or oviparous, he had accepted the growing indications of Owen's ovoviviparous view. Thus, concentrated on collecting platypus uteri to support the concept of the young hatching internally within the mother, his research approach held a fatal flaw: it failed to entertain and hence search out any evidence of laid eggs in the nesting burrows.

There were other important factors. Bennett's resistance to information supplied by the Aborigines diverted him from the real facts. A variety of colonists and bushmen repeated the Aborigines' testimony that the platypus laid eggs. But after his early attempts to question them, Bennett stuck to a misguided view that no dependence could be placed on 'native accounts' and that naturalists 'must seek for information in their own investigations'.

In 1871, after many independent ventures himself, Bennett enlisted his son, George Frederick Bennett, in the hunt. A novice in natural history, young George had moved to work on a pastoral property in southern Queensland. Keen to prove himself to his father, and perhaps to blaze his name across Australia's zoological science, George attacked the burrows. Opening two on the Lockyer River in October 1876, in one he found two young platypuses which he judged to be three weeks old, 'rolled up like a ball'. They measured 13 centimetres from the back to the tip of their tail and 'were very broad and fat'. A month later, excavating a second burrow, he captured two infants, which—following his father's custom—he tried to keep alive for several days on milk and sugar with the usual lack of success.

George Bennett senior had sketched out a blueprint to guide his son's enquiries. Its wish-list included the season and manner of copulation, the likely periods of gestation, the physical nature of the structures developed for the support of the foetus during gestation, the exact size, condition and powers of the young at the time of their birth, their method of suckling, the period of 'the lacteal nourishment', and the age at which the animal attained its full size. They were the very items that Owen had set down as the principal points to be determined in his article on the monotremes in the *Cyclopaedia* over 30 years before. There was no mention of an egg.

After several bouts of killing and pickling, young George was in direct communication with Owen. The subject, he enthused, was 'most interesting' and grew more and more so as one advanced in it. Owen offered a modest cover of costs and clearly expected a good return from the new recruit.

'There is still the impregnated womb of a platypus or echidna to be got', he wrote jauntily to the New Zealand geologist, Julius Haast, in 1878, 'of which a young correspondent in Queensland is in earnest quest'.

Bennett, young George and Owen himself, however, were to be pipped at the post.

In September 1883, a young Scottish embryologist from Cambridge University, William Caldwell, the holder of the university's newly established Balfour travelling studentship, arrived in Australia with several major research agendas in his knapsack. His plans related broadly to 'the development of the peculiar Australian Mammalia and the *Ceratodus*'— subjects that were first broached with him by his professor, the renowned embryologist F. M. Balfour after whom the memorial scholarship (following his untimely death by a fall in the Alps) was endowed and named.

Born in 1859, Caldwell belonged to the new guard of young biologists trained in embryological studies and reared on the works of Darwin and Huxley. His special research topic was species reproduction, their survival and living form. Colleagues hence looked to the brilliant 23-year-old postgraduate to transcend Richard Owen's old morphological thinking and expose the reproductive physiology of the two 'living fossils', the platypus and the remarkable lungfish, *Ceratodus*, to scientific view.

Generously financed with additional grants for equipment and expenses, Caldwell landed in Sydney with a sheaf of letters from the Colonial Office and the Royal Society of London to the colonial governors of New South Wales, Victoria and Queensland. Given rooms by a receptive government in Sydney, he was soon in touch with key colonial scientists.

Young Caldwell was a professional. After devoting time to one part of his research agenda—the early post-natal development of the Australian marsupials—and working on the kangaroo, possum and koala in northern New South Wales, he moved in April 1884 to the Burnett River in southern Queensland where the *Ceratodus* was to be found. Another Antipodean conundrum that taxed biological understanding, the *Ceratodus* (known subsequently as *Neoceratodus*) was a living representative of a distinct order of fish which in an adult stage had the form of a fish with gills but also possessed lungs. The *Lepidosiren* found in the Amazon and the *Protopterus* in certain rivers of Africa were of related kind.

The embryology of this bizarre creature was a key object of Caldwell's three-pronged Australian research. Camped on the Burnett River, assisted by some 50 Aborigines living nearby, he trawled along the river and its many waterholes for four active months until he found a trace of *Ceratodus* eggs laid well hidden among the weeds. Finding at last some eggs 'covered in a quantity of gelatinous matter', the young embryologist set about rearing them until they were identical with the adult fish with its singular gills, lungs, scales and the rudimentary legs that propelled it into trees and weeds.

Across the same period, the resourceful Caldwell also deployed some of his large corps of Aborigines to hunt out the echidna and platypus.

He had already made a dramatic onslaught on the platypus

the previous October through December in the chilly district of New England, New South Wales, while conducting his work on the kangaroo. In one lagoon, he told a surprised Dr Bennett, he had killed 'seventy female Platypi [sic]' which he examined carefully for impregnation without result. He would apply the same kill and purge tactics on the Burnett.

The Burnett River was known both as a good platypus habitat and as a locality, becoming rarer in the eastern Colonies, still occupied by a large number of Aborigines. Rounding up an Indigenous army that ran to 150 in the monotreme breeding season, Caldwell was soon applying himself to the echidnas and 'segmenting ova from the[ir] uterus'. In the third week

Joseph Lycett, 'Aborigines hunting'. Large corps of Aborigines were used by Caldwell on the Burnett River, Queensland to assist him in solving the platypus mystery. 'Without the services of these people,' he acknowledged, 'I should have had little chance of success'.

of August, holding the animal upside down, a surprised Caldwell dropped 'the laid eggs' from the pouch of an echidna.

Culinary factors favoured his scientific search. The echidnas, Caldwell found, were the Aborigines' favourite food, but 'it was only occasionally, and then with great difficulty, that I persuaded them to dig for *Ornithorhynchus*'. Both Aborigines and their dogs refused to eat the small aquatic animal.

Nourishing or not, *Ornithorhynchus* was on Caldwell's scientific menu, and by the second week of August, he was finding similar stages in the ova of the platypus to that of the echidna. In the third week, he hit the jackpot. On 24 August 1884, the young Scot shot a platypus 'whose first egg', he reported in a statement that made biological history, 'had been laid; her second egg was in a partially dilated *os uteri*' [mouth of the uterus].

Five days later on 29 August, within reach of a country telegraph station, the excited scientist communicated the significant news to the Dean of Science at Sydney University, Professor Archibald Liversidge, requesting that he forward it to the British Association for the Advancement of Science meeting (for the first time outside Britain) in Montreal.

The die was cast. There before the assembled scientists drawn from all parts of the British Empire, Caldwell's terse but telling words rang out: *'Monotremes oviparous, ovum meroblastic'* which, being interpreted, meant that monotremes lay eggs; the soft-shell eggs have large yolks which do not divide into cells but are absorbed as food by the developing young, as with birds.

To the biologist, the term 'meroblastic' indicates a mode of cleavage or initial division into embryonic cells after fertilisation. A yolky egg shows incomplete cleavage. Egg-laying land vertebrates, reptiles and birds tend to produce yolky egg cells with meroblastic cleavage. Most mammals, however, show complete or holoblastic cleavage. Caldwell was hence trumpeting in his message: these curious mammals had a reptilian character.

The long, closely guarded secret of the monotremes was out. The silence was broken. 'No more important message in a scientific sense' said the President of the BAAS's Biological Section, Professor Moseley, reading the transmitted message to the assembled company on 2 September 1884, 'had ever passed through the submarine cables'. The monotremes, he declared, now evidenced their clear intermediary link between reptiles and mammals.

The prize had gone to the Scotsman. Wisely, Caldwell used the telegraph again, this time to inform George Bennett in Sydney that he had 'obtained all stages development Monotremata oviparous Meroblastic'.

Yet, in one of those coincidences that mark discovery in science, the very day of Caldwell's discovery, across Australia near Adelaide, the German naturalist Wilhelm Haacke, curator of the South Australian Museum, found the remains of an egg in the pouch of an echidna. Unaware of Caldwell's tidings circling the globe, he eagerly exhibited his discovery at the Royal Society of South Australia on 2 September—the same day as the Montreal meeting.

But it was Caldwell who scored the triumph. He had applied a strategic tactic of search and destroy with a research method of dedicated concentration to the problem that, through speculative theory, interruptions, delays, muddle and differing and random approaches had baffled science for 90 years.

When he came to Australia, the young biologist revealed, he held a strong belief that the platypus and the echidna produced their young 'in much the same way as marsupials', by live birth. In his first months in the country, he thought he had the proof. Not so, he admitted. Presenting his findings to a meeting of assorted greybeards of the New South Wales Royal Society in December 1884 with his diagrams before him, he related how he had found a cellular membrane in the uterus of his 'discovery' platypus which was probably part of a ruptured egg. Using his drawings, he went on to describe the large food yolk of the platypus egg and its meroblastic segmentation.

The difference between the Australian marsupials and the higher mammals of the Old World—the dogs, cats and sheep —the young Scotsman told the colonial naturalists, with perhaps a touch of condescension—was that the marsupial young were born in a very early stage of development, and that the embryo in the uterus before birth had no vascular attachment to the walls. There was, hence, unlike the higher mammals, no blood nourishment passing from the parent to the young animal. Yet both marsupials and higher mammals possessed a very small amount of food yolk. The monotremes, with their soft-shelled eggs of large food yolk, were singularly different.

During his Burnett research, Caldwell was already aware of the presence of platypus eggs, for, as he recounted, he invariably found that the female platypus had two eggs while the echidna produced only one. In both, the eggs with their shell were about the same size—15 millimetres long and 12 millimetres wide. But the vital discovery of the platypus's egg in a female that had produced one and had the other on its way through the passage was, said Caldwell, 'a lucky chance'—a chance that enabled him to pinpoint the time when the eggs were laid.

Beaten to the post, George Bennett showed generosity. 'The conclusions arrived at by so excellent an embryologist', he was at once in touch with Owen, 'is entitled to be regarded as trustworthy'. But his chagrin was plain. 'My anticipation has therefore not been realized as I have always considered they would prove to be ovoviviparous.' But wait and see was Bennett's message to his old friend. Caldwell's development of the mystery will be, he wrote, of the greatest interest to many 'but to none more than to ourselves that the problems of the Monotremes will be solved during our lifetime'.

Both Bennett and Owen were old men: Bennett now 80, Owen in his 79th year, when the mystery was solved. Their labours had stretched over half a century. Indeed, Bennett and his son's consignments of echidnas and Owen's ensuing paper on the ovum of the echidna reached to the very moment of Caldwell's finds. If events had conspired against them, Bennett, the part-time naturalist, as historian Jacob Gruber observed, was working against the dictates of the new professionalism of his time.

Bennett himself perceived the professional at work. 'There is a great difference', he wrote Owen later that year,

in pursuing these investigations of the generation of the Monotremes at long intervals of time and being enabled to camp out for nearly six months about their haunts, submitting them to daily investigations, which has been done by Mr Caldwell. This is the only way, when it can be done, to achieve success, and Mr Caldwell has a still larger advantage over the amateur Naturalists in being a well educated scientific embryologist with all his attention and knowledge devoted to the investigation.

Yet humanly, after his long vigils, Bennett felt some irritation with the platypus. 'Who would have thought', he reflected, 'that an animal with so large a milk gland should actually demean itself by laying small white eggs'.

Owen's recorded thoughts on the matter have not survived. It is doubtful that they would have been as kind.

For Caldwell, fortune favoured the prepared mind. He had broken the paradigm. Richard Owen's long dominance over Australian biological science was at an end. Caldwell had resolved a scientific mystery that had plagued zoology for 90 years. He would brook no dispute from ovoviviparous claimants in the Colonies. The outcome of his investigations, he told his audience of separate creationists at the Royal Society of New South Wales, were facts, not theories. He wanted no letters or discussion; as facts they could not be argued. As an evolutionist who recognised that each living form had descended from some 'differently constructed ancestor', he was the first, in the wake of Darwin, to attempt to fit the monotremes into the evolutionary frame.

In places like the Burnett River, where the platypus had formerly lived out its peaceful existence, solving the mystery had its price. Science had advanced over the massed bodies of the monotremes. Bennett and his son were vigorous contributors to the carnage. But their killings paled beside the tally of the thousands sacrificed by Caldwell and his Aboriginal team. Caldwell, moreover, returned to England with over 1300 echidna specimens for his future work.

But, as Huxley remarked, science marched on the solid grounds of observation. Twelve thousand miles from the centres of hypothesis and theory, the platypus mystery was unravelled in Australia. Significantly, her oldest inhabitants, the Aboriginal people, earlier considered unreliable informants —played a timely part. 'Without the services of these people', Caldwell acknowledged, 'I should have had little chance of success'.

12

THE NEW
MEN

You have so many pretty things to show and so
much to say that people in England would like
to hear.

Grafton Elliot Smith to J. T. Wilson,
February 1904, in Patricia Morison,
*J. T. Wilson and the Fraternity of
Duckmaloi*, 1997

William Caldwell's discoveries heralded the new importance of embryology in scientific research. The study
of the embryo or ova before the offspring's birth was emerging
as a crucial scientific field. Lecturing on it at Cambridge,
Caldwell's professor, F. M. Balfour, had pushed the subject
forward with his *A Treatise on Comparative Embryology* published
shortly before his early death, and the discipline gained ground
in Britain and the USA. Darwin too had commended it as an
aid to determining lines of descent from a common progenitor.

Caldwell, the inheritor of both the study and Darwin's

view, was uniquely placed to carry the discipline forward. But, praised in biological circles on his return to England, the young virtuoso delivered his introductory paper on the embryology of Monotremata and Marsupialia to the Royal Society of London and promptly joined the 'brain drain'. He abandoned his scientific career to become a successful paper manufacturer in Scotland and drifted off the scientific page.

He was, of course, well aware of just how much on the edge of knowledge his remarkable discoveries were. Years, he rightly predicted, would elapse before the details of the development of the platypus and echidna could be extended and interpreted.

That the platypus was an egg-laying mammal was now accepted, although *The Illustrated London Magazine* was still blithely referring to 'the egg-laying myth' in its columns in 1886. Relaying Caldwell's message in Montreal, Professor Moseley thought the platypus evoked a reptilian ancestor, but its descent remained obscure. Writing his *Mammalian Descent* a year after Caldwell's revelation, the vertebrate authority Professor W. K. Parker allowed that he was still 'at a loss' as to how to deal with this complex animal:

Here is a beast—a primary kind of beast ... whose general structure puts it somewhere on the same level as low reptiles, and old sorts of birds; but in which there are characters much more archaic than anything seen in Serpents, Lizards, Tortoises, Crocodiles, or in Emus. Therefore the existing reptiles and birds must stand aside as having nothing to do with the family tree of the Monotremes, although in some things they are like these beasts, and many of their organs are formed in a similar pattern; they are all equally below the morphological level of the nobler Mammalia.

But science is the product of many players, of untied ends and of new beginnings. The torch flung down by Caldwell would shortly be taken up by others. For even as Caldwell brought together his descriptive paper for London's Royal Society in early 1887, one sharp student from the Medical School of Edinburgh University—another Scot—James Thomas Wilson, was on his way to take up a demonstrator's post in anatomy in the new Medical School of the University of Sydney.

Wilson was 27, a mere eight years younger than the university itself. Within three years, rising fast to became the first Challis Professor of Anatomy, the large and forthright 'Jummie' was well placed to gather round him a small circle of talented young biologists to join him in important Australian research.

These men were the new breed, Darwinians all. Wilson himself, trained in anatomy and embryology; young J. P. (Peter) Hill, a student from the Royal College of Science in London; physiologist and pathologist Charles Martin, a former lecturer in anatomy at King's College, London—all were professional scientists making their way to high places via posts in the Australian Colonies. The triumvirate drew in an outstanding local medical graduate, Grafton Elliot Smith, and were soon ready to make the indigenous fauna their research arena and to fit those most curious species, the monotremes and marsupials, into the genealogical tree of life.

Caldwell had determined one fact; there were countless more to be discovered. His 1887–89 papers had fuelled Wilson's interest. But Wilson, like others who followed, found that the material collected by Caldwell and his scientific exposition was disappointing. Trained in histology (the study of tissue) and microscopy, the very 'instruments now at the command of embryologists' which Caldwell's departing figure had extolled, Wilson and his colleagues would shift the centre of monotreme research to Australia.

Moving casually about the countryside in their free time from teaching, they took their moniker, 'the Fraternity of Duckmaloi', from their favourite hunting ground for the platypus on the gumtree-fringed Duckmaloi River which flows on the inland side of the Great Dividing Range, near Oberon, New South Wales. Using specimens captured from its banks and from rivers running on the coastal side of the range, they bent their individual and collective minds to amassing a spread of knowledge on the physiology of the platypus. Their research probed the duckbill snout, the reptilian shoulder girdle, the skull and brain, the dumb-bell shaped bone or 'os paradoxum' in the skull (the bony nodule that furnished the embryonic monotreme with a 'shell-breaker') and the hind spur on the male—and built understanding of the neurology, morphology and evolutionary physiology of the monotremes.

For all their energy and enthusiasm, platypus hunting proved a daunting task. As their predecessors knew, the animals were hard to find. There was also evidence that fur hunters had reduced their numbers in New South Wales. Again and again the animal's amazing construction skills defied the 'Fraternity'. Their sorties on the long labyrinthine burrows—built some 30 to 46 centimetres below the ground, up to 18 metres long and securely plugged at both ends to conceal the nesting chamber—often yielded no other reward than companionable evenings spent around a campfire on the chilly August and September nights.

Wilson and Hill usually captured up to six males to every female but in 1894 four female platypuses were shot at Wellington, three having just laid their eggs and the fourth with two eggs in her left uterus ready to be laid. The following year the two men used the Colony's Linnean Society journal to publish the first precise description of the structure of a platypus embryo and its foetal membranes. It became a foundation paper in the field.

Wilson and Hill would concentrate their efforts in this new and exacting area of research. With a dozen suitable

Members of the 'Fraternity', J. T. Wilson, J. P. Hill and G. Elliot Smith, platypus hunting on the Duckmaloi River, September 1895.

specimens finally at their disposal, they turned to the technology of the microscope and to the skills of embryological dissection and reconstitution to reveal the composition of the developing platypus egg.

After Caldwell's 'lucky chance' discovery of one egg with another in the female's passage about to be laid, the gestation period of the platypus was still unknown. With hard-won eggs now available for detailed embryological scrutiny, the researchers set out to make delicate slide sections of the egg in its various stages of development.

Their intricate work brought results. Working across the years 1902–06, the two colleagues traced the stages of the development of yolk and membrane to the point where the egg was about to be laid. Their findings were explicit. The elusive platypus egg proved both similar to and distinct from that of birds. The platypus ovum began at the time of fertilisation as a small yellow sphere of about 3 millimetres in diameter. Fertilised, it became surrounded by a very thin layer of albumen (the white of the egg) and outside this, by a thin, transparent parchment-like shell.

In birds the albumen is deposited in several layers in which the yolk is suspended in the middle of the egg. Membranes attach the yolk to a rigid calcareous shell. The difference between the egg of a bird and that of a monotreme, Wilson and Hill discovered, was that in the bird the shell was deposited around the fully formed egg and there was no increase in size during the growth of the embryo. But in the platypus (and echidna) the egg shell increased in size and altered in structure along the spine or axis of the embryo during its intra-uterine development. The thin shell stretched and thickened from an initial 4 millimetres in diameter to a spherical size of some 10 millimetres. It could expand to a maximum of 16–18 millimetres in length and 14–15 millimetres in width by the time of its being laid.

Only the monotremes shared this distinctive pattern of embryonic growth. Why? The researchers' answer lay in the fact that there was insufficient nutrient in the yolk-mass of the fertilised ovum to produce the young platypus. The foetus was thus nourished by a nutrient fluid flowing to it from the uterine wall which enlarged the egg and progressively disrupted and liquefied the yolk.

Wilson and Hill's conclusions broke new ground. In the earliest stages the development of the platypus, their findings ran, was similar to the 'meroblastic' (yolk-filled) ova of reptiles and birds. However, nourishment of the embryo inside the egg from the uterine wall was a mammalian feature and did not occur in birds and reptiles. The Australian duo had, they believed, discovered the link between reptiles and mammals and the origin of mammals.

Their research dazzled biologists around the world. By 1903, their careful observations were judged in Britain 'the acme of platypus work'. In 1905, on sabbatical leave in Europe and Britain, Wilson displayed the revealing stereophotographs of embryological material of platypus development and monotreme reproduction at the first international congress of anatomists in Geneva. They had, as Wilson's biographer notes, become the foremost authority on monotremes.

The work Wilson and Hill conducted on the platypus brain was also far reaching. By examining early neural development in the embryo, they found that the platypus brain is not formed into three parts like other mammals but into two, a hind part and a fore part, and it is under the fore part that the primitive spine terminates. They also identified an enormous growth of cells (relative to the rest of the neural system) on each side of the flat forebrain which they found to be continuous with the brain.

Charles Martin and Grafton Elliot Smith added other discoveries. Lecturing at Sydney University and later at Melbourne University, Martin conducted definitive work on the heat regulating mechanisms of monotremes and researched the spur of the male platypus's hind limb. This feature— a subject that had engaged Bennett, Owen, Gould and many a bush settler—he concluded, transmitted a poisonous secretion from the duct of the femoral gland, but was only used in so peaceable an animal as a weapon against other males during the mating season. The brilliant young Elliot Smith centred research on the platypus brain in a study which, extending widely to marsupials, reptiles and other mammals, propelled him to a distinguished career as a leading neuroanatomist in Egypt and Britain.

Hill, attracting international recognition for his innovative work, left Sydney in 1906 to take up successive chairs of zoology and comparative anatomy in London, where he continued his intensive study of monotreme egg development until well after his retirement in 1938. During the 1930s, he collaborated with a former student, Thomas Flynn, appointed to the Chair of Biology at Belfast, who had arrived in London with a rich collection of monotreme eggs and embryos. Their detailed research shed further light on the development in the monotreme egg that bridges the gap between reptilian and avian development. Hill's continuing work also established that monotremes, marsupials and placental mammals such as the cat marked specific evolutionary stages through which the higher mammals have all passed. It won for him the Royal Society's high award of the Darwin Medal.

Wilson clarified other uncertainties. Focusing with Martin on the platypus snout, he determined that it was not—as earlier investigators with access only to dried or preserved specimens had suggested—'horny' or 'leathery', but cartilaginous like the pre-nasal structure of the pig. Further fundamental research on the snout revealed a complex nervous system, peripheral sensibilities and remarkable touch receptor functions.

A prominent figure in Sydney, Wilson was wooed back to Britain in 1920 to take up the Chair of Anatomy at Cambridge University.

The 'Fraternity of Duckmaloi' had changed the face of scientific monotreme investigation. All collected fellowships of the Royal Society for their contributions, and all rose to prominent positions in Britain's scientific institutions. Most importantly, these 'new men' transferred the major site of monotreme research from Britain and Europe and anchored it in Australia. In doing so, they, far more than Caldwell, stripped the veil from monotreme reproduction and opened early mammalian development to scientific view.

'THE PLATYPUS MAN'

O! Thou prehistoric link,
kin to beaver, rooster, skink,
Duck, mole, adder, monkey, fox,
Palaeozoic paradox!

Harry Burrell, 'The Mud-Sucking
Platypus'

New men come in different guises. While the Fraternity of Duckmaloi were pinning down the anatomy and embryology of the platypus and communicating their academic discoveries to scientific circles around the world, a young knockabout lad, Harry Burrell, was out along the Namoi and Manilla rivers of New South Wales cocking an inquisitive eye at the curious aquatic animal and putting together a store of information on its behaviour and habitat.

Born in Sydney in 1873, Burrell had only an elementary education and had embarked on a wandering life, including a spell as a vaudeville comedian, before he settled with his bride in the first year of the new century on a grazing property near Manilla. There he soon established a small zoological garden in which he kept birds and one or two marsupials. And, while out and about collecting water-weeds for his ducks, he met his first platypus, and fell in love.

A practical observing naturalist of the old school, the Harry Burrell who gazes from his photograph is a tall lean man with a 'get-up-and-go' air, who, across an active 25 years, brought together and made widely accessible a wealth of old and new data in his classic work *The Platypus* which he published in 1925. For nature lovers, zoologists and generations of fascinated readers, Burrell became a household word.

Burrell's platypus of rivers, streams and land was not romanticised. Drawn from prolonged observation of its living form, there was something compellingly sure and credible about his accounts which linked the engaging animal's physical appearance and capacities with its daily life.

> The platypus is an aquatic animal, but its bodily form shows no very marked adaptation to aquatic conditions. During countless ages of life in the water it has not developed the 'stream-lines' of porpoises and seals. Its body is squat, clumsy, and reptilian, as are its short, thick limbs. The adaptations to aquatic life are mainly two: the enormous webbing of the fore-foot or paw, which is the actual swimming organ, and the flattened tail, which is used as a rudder and helps in diving ... It is not a rapid swimmer, nor an especially graceful one.

The face was unveiled:

The small, bright eyes are remarkable only for their position high up on the head. Since they are not used for vision under water, they have come to be placed where they will be of most service to the animal as it floats at the surface, munching its catch. The absence of an external ear would seem to be a primitive character; in its place there is a curious modification of the orifice of the auditory meatus. This orifice lies at the posterior end of a facial furrow, the eye lying at the anterior end, while the furrow is incompletely divided into two by an oblique fold of skin. The edges of this furrow act as a long pair of lids, by means of which both eye and ear may be tightly closed at the will of the animal. The aural aperture can also be dilated and contracted while the eyes are open, and can be 'cocked' to catch sound. The arrangement of the lids serves also to keep out water while the animal is submerged, and earth while it is engaged in burrowing.

The duck-bill, snout, beak, muzzle—all the names which differing writers and theorists had deployed across the years—was sketched plainly for the reader's eye:

The muzzle, which shows some resemblance to a duck's bill in the dry condition, is very different in the living animal. The naked skin is thick, but soft, moist, and flexible, very unlike the horny beak of a bird. On the upper surface it is slate coloured; on the under, of a yellowish flesh-colour, often broadly mottled with greenish slate.

From the base of each mandible a cuticular flap projects backwards over the fur of forehead and throat.

The muzzle with its flaps, is a highly specialised sense-organ. The whole of its exposed surface, both above and below, is pitted with minute pores, which extend on to the cuticular flaps and mark the sites of the highly specialized touch-corpuscles. When under water, the animal depends principally on its delicate sense of touch for finding its way about ... Even when in the open air, it probably depends largely upon the muzzle, since its eyes are so placed that it cannot easily see objects straight in front of it on the ground ... It is well known that it can squeeze through very narrow spaces, and it is possible that the flaps are used as a gauge by means of which it can tell whether it is safe to go on, or wiser to withdraw. Whether this be so or not, the flaps are assuredly a part of the great tactile organ, and not a mechanical shield. In some of the earlier figures, drawn from dried skins, these flaps are shown standing up at right angles from the surface of the head; in the living animal they are always laid back upon the fur.

Burrell when he wrote had access to Bennett and Caldwell's accounts and he nods to the scientific findings of Wilson and Hill. But much experiential knowledge flowed from his own perception and examination of an animal that held him closely in its thrall.

His observational fieldwork in creeks and rivers was supplemented by his growing skills in keeping the platypus in captivity. Alert to their habits, he devised an enclosure—a roofed-over brick structure, floored with a thick layer of river earth, representing the burrow, and connected by a submerged tunnel to a wire netting-enclosed cement pond, which resembled a kind of suburban platypus habitat.

At first, in 1910, his efforts were experimental. Finding a female platypus in a river trap, he placed her in the enclosure and shortly after added two more females, and later, to the noisy annoyance of the sorority, two males. It was hardly a pushover. Platypuses are monumental eaters. Burrell, after working vigorously for six hours daily with mattock and shrimping net to capture a diet of earthworms, freshwater shrimps, larvae of scarab beetles and pond snails, managed to collect an inadequate two pounds of food a day. One by one the captive animals died, while the surviving female, comfortably and instructively ate the amount of live food thought adequate for five platypuses. After nine weeks in captivity, having built her strength for departure, she escaped by tearing her way through the pond's wire netting.

In Manilla, and later Sydney, Burrell's knowledge of the platypus grew. Highly selective in eating, its natural food resembled that of a bird rather than of a mammal, he wrote, and 'it will starve to death in the presence of food which no longer pleases it'. It was, too, 'impatient of observation, and resents being handled' and was 'easily killed by too much excitement'.

In comparative studies, he found the echidna 'a dull animal' despite the fact it had the higher mammalian convoluted brain while that of the platypus was smooth. Yet there was nothing stupid about the platypus. The brain, Burrell reported, was 'surprisingly large', definitely mammalian in structure, with a cerebrum better organised than that of the lower marsupials, and a large surface of cerebral cortex that denoted a considerable intelligence. It was a point that a few years earlier had fascinated the South Australian anatomist Frederick Wood Jones. The platypus brain, Jones noticed, 'though built upon a lowly plan, possesses a great share of specialised cerebral cortex or grey matter'.

Burrell offered a prescient rider. From his long observation of the platypus in the wild and of its convoluted burrows with their carefully constructed side tracks, its acute response to warnings in its natural environment, and the successful attempts of the captured platypus to return to its home river, he contended that the platypus possessed a 'sixth sense'.

Importantly, he was also able to shed further light and offer some robust deductions on the animal's breeding and egg-laying habits. After platypus mating, an event the assiduous Burrell had managed to witness only twice across a watchful 25 years, the female usually left her accustomed feeding-grounds and, given to solitude, selected an isolated stretch of bank suitable for nesting—one platypus stationed carefully with some 5 kilometres of river at her exclusive command. The nesting burrow was not the same as that used for normal habitation. It was more elaborate. Its entrance was shaped like a low archway after which the narrow passage led off in its sinuous way, side-tracks offering a diversion, until it ended in the large oval nesting chamber where leaves, dry grass and tree roots made up the nest. Such burrows were abandoned as soon as the young could shift for themselves.

Fossicking around the river bends, Burrell noted that the earliest time he took eggs from a nest in the breeding season was in late August. On the same day in August 1925, he unearthed twin young, which he judged to be three days old, from a neighbouring burrow. He considered then that the date of the first egg's laying was probably the first week of August. Watchful surmise was his rule of thumb. He showed foresight. The exact period of platypus gestation is still poorly understood today.

Burrell's chapters were rich in detail. His practical eye picked up the information that neither nineteenth-century naturalists nor embryologists on the cusp of two centuries had divined.

His further examinations revealed that when the platypus's two eggs were laid, they were invariably found joined together side by side. Lying behind each other in the left uterus before laying, separated in the laying by an appreciable interval, each laid egg was coated with a thick sticky secretion from the oviduct walls.

While the maternal platypus's carefully plugged burrow protected the birth process from eyewitness view, Burrell reasoned that, curled in the nursing nest, tail between her legs, she dropped her eggs from her cloaca into her soft rubber-like hands and held them there until each egg capsule, in contact with the other, became fixed together. These she subsequently clamped to her warm abdomen.

The incubation process also needed imaginative projection. Burrell believed that the mother incubated the eggs by curling about them and placing them between her abdomen and upturned tail, a position in which he had once found a female platypus holding her minuscule young.

He expressed the process aptly in his verse, 'The Mud-Sucking Platypus':

Epipubic bones support
Dimpled abdomen; in short,
In that slight depression she
Incubates her progeny
Warmth increased for eggs and young
By her tail, well underslung;
Snugly cuddled to her breast,
Mother nature does the rest.

The infant platypus after hatching, Burrell found, unlike other mammals, was not suckled by the mother for several days—a finding he confirmed by extracting from the burrows and examining several mothers in various stages of lacteal dryness. Rather, the female's 'primitive' milk glands remained dry until excited by the mechanical stimulus of the movements of the young.

Much new evidence emerged from Burrell's work. He consolidated knowledge and shared his growing intimacy with an animal which, in essence, rejected the intimacy of man. But his own judgment of his achievement was modest. 'The platypus', his book concluded, 'is, from a scientific point of view, perhaps the most important mammal that exists, and a great deal of anatomical investigation remains to be done'.

But, if intimacy defeated him, Burrell's attempt at closeness led him to the ingenious invention of a portable artificial habitat for the platypus which he named the 'Platypusary'. A vast refinement on his original 'suburban' platypus habitat, it consisted of a tank to represent the river, a labyrinth to simulate the burrow on the bank and a sheet metal tunnel connecting both. By devising sloping runways and several

Harry Burrell with the portable 'Platypusary' he designed.

narrow-fronted apertures through which the animal had to pass to the burrow from the pond, he ensured that it arrived almost dry in its sleeping quarters. This time, he made the floor not of mud, which sullied the pond, but of sand and shell grit as no platypus tolerates water that is unclean.

With a 'Platypusary' at his disposal, Burrell was able to display the animal at the Sydney Zoological Gardens in 1910 and to provide the tutelage for the dispatch of five male platypuses by sea and rail to New York in 1922, the one survivor of the journey becoming the first live platypus to be seen—by a stampeding crowd—outside Australia.

'The spell of ten thousand years has been broken,' exclaimed the Director of the New York Zoo upon his arrival there. 'The most wonderful of all living mammals has been carried alive from its insular confines of its too-far-distant native land, and introduced abroad.' He was slightly disconcerted, however, that an animal so small (for it was half-grown) 'could chamber a food-supply so large'!

The country boy turned independent zoologist, the 'pioneer without peer' as colleagues knew him, liked to be known simply as 'the Platypus man'. He is commemorated as an 'Unsung Hero of the Outback' in the Australian Stockman's Hall of Fame at Longreach, Queensland.

Burrell's know-how and skill in handling the platypus in captivity passed on to a younger contemporary, David Fleay. Born in 1907 and trained in science at Melbourne University, Fleay was also a lively observer of nature in the wild from boyhood. Appointed curator of Australian animals at the Melbourne Zoo in 1934, in 1937 he became Director of the Healesville Sanctuary established in a secluded rural region of Victoria outside Melbourne. In that first year he installed a 'Platypusary'.

Its first inhabitant was 'Jill', a tiny platypus that had wandered prematurely from a nesting burrow and was found shuffling down the middle of a country road. Two years later, as flames enveloped the ranges of Victoria in the devastating summer bushfires of 1939, 'Jack', a half-grown male platypus picked up by the beam of a torchlight one evening, was plucked from a shrinking pool in Badgery Creek. Fleay's success in domesticating and rearing these delicate and unpredictable creatures in captivity was news that flashed

David Fleay with Jill, the mother platypus (right), and Corrie (left), the first platypus to be born in captivity.

around Australia and overseas. Four years later, he was called on to perform the complex task of preparing and shipping three living platypuses to the New York Zoo.

Sanctioned by the Fisheries and Game Department of the Victorian Government as a scientific and educational mission, the long planning got under way and a hunt for suitable candidates was begun in neighbouring Badger Creek. From a catch of nineteen platypuses, three were selected—two females, 'Betty' and 'Penelope', and 'Cecil', a placid male and placed in portable platypusaries. After a sequence of adverse reactions, adjustments and delays during which the animals gradually grew accustomed to appearing before the public, the arduous journey was launched.

In March 1947, Fleay and the three platypuses flew to Brisbane and were met by the *Pioneer Glen* from Melbourne, carrying food supplies of 7000 frozen yabbies, 136 000 frozen worms, 22 000 live grubs, 23 000 live worms, 45 live frogs and a ready supply of duck and hen eggs to make a platypus custard for their voyage across the Pacific.

Every accommodation was made on the journey to keep the temperamental animals healthy and undisturbed and ready to emerge from their burrows for their evening swim in the platypusary tanks. A sympathetic captain suspended painting and deck chipping, turned off flood-lights, placed vertical baffle boards to prevent the platypus tank water from slopping, muted the fog-horn at night and even changed course to avoid rough weather in the interests of platypus harmony. When food supplies ran low, Pitcairn Islanders, alerted, waited expectantly in their light boats offshore for the *Pioneer Glen* and its 'platypus zoo' with kerosene tins of worms. More live worms were flown from New York to reach the ship at Balboa.

At length, on 25 April 1947, a triumphant David Fleay landed with the three platypuses at Boston for the flight to New York. It was the arrival of the first female platypus on American soil. Three days later, the preview exhibition of the new arrivals for members of the New York Zoological Society was ceremoniously opened by the Australian Ambassador to the United States, Norman Makin. The American and

Australian flags waved proudly in the cool sunshine. The Director of the Zoological Gardens, Fairfield Osborn, wrote in the bulletin of the Zoological Society, *Animal Kingdom*, of the extraordinary wonder and excitement that greeted the animals' arrival in the US exactly a century and a half after its first appearance in Britain.

> The day has finally come, the first that a generation of American eyes could look upon the 'most marvellous of living mammals'. On the day of their arrival batteries of photographers from newspapers and newsreels made an array such as that which would greet the most distinguished of foreign visitors. Since then, the radio, magazines and the press are all 'gone platypus'. We were sure that they would take America by storm. So they have.

There was, said Osborn, perhaps 'something whimsical about the whole matter'. Millions of years ago, the platypus had 'had the choice of climbing the broad path of evolution to the upper levels of the mammalian hierarchy, but . . . they continued as they were. So they still lay an egg, still suckle their young, and so remain the living link between the misty eons of the past and this very bright summer day in 1947.'

Fleay, the zoologist who had spent 470 days of preparation for this event and much trial and anxiety, was more practical. 'In the twentieth century', he summed up, 'the fame of the amazing platypus has spread world-wide, far beyond scientific circles . . . but don't underrate the difficulties of pioneers facing the eccentricities of *Ornithorhynchus*! Never did any furred animal offer a greater challenge to closer acquaintance.'

14

THE PLATYPUS
GOES TO WAR

...having expressed himself in writing clearly
and fully about a subject [Churchill] was inclined
to think that the matter was accomplished,
largely done. He was not unaware of the obstacles
of bureaucracy...a special label affixed to papers
and directives marked 'Action This Day'.

John Lukacs, *Five Days in London*,
May 1940, 1999

David Fleay's success with Jack and Jill at Healesville since
the late 1930s was widely known before the adventurous
exploit to America in 1947. At the height of war, British Prime
Minister Winston Churchill was clearly aware of it.

As 1943 dawned, developments in World War II challenged
on every front. Churchill and US President Franklin Roosevelt
met at Casablanca in February and Churchill continued on
to Cairo to overview crucial military strategy. Rommel, the
German 'desert fox', was fighting the British Eighth Army
and the Americans in Tunisia. Field-Marshal Montgomery was

'sharpening his claws' for the final Allied battle against the Germans in North Africa. With British troops stretched out from Gibraltar across the Mediterranean to India and into Africa, there was strong pressure from Stalin for a second front. Churchill, juggling his multifarious duties as Prime Minister and Minister for Defence—overseeing strategy, planning military and civilian efforts, determining the critical allocation of resources between the Home Front and the many theatres of war—turned his lively mind to another subject.

At the end of February 1943, Churchill cabled the Australian Prime Minister, John Curtin, with an unusual request. Would he send six live platypuses as soon as possible to the United Kingdom? Churchill was already an admirer of the grace and beauty of Australia's black swans—half a dozen of these, presented to him in the 1930s, swam contentedly about the lake at his country home, Chartwell. But news had reached him of the two platypuses, Jack and Jill.

At Healesville, David Fleay was hence vastly surprised in March 1943 to receive a visit from two hatted officials of the Commonwealth Department of Health where Churchill's request had been conveyed, bearing news of this proposed wartime mission. It was to be 'hush hush', with no preceding correspondence to explain it, and orders for a total ban on any media coverage. 'It was', Fleay admitted, 'the shock of a lifetime'.

Whether it was to be interpreted as a good boost to British morale or a splendid propaganda diversion, 'we never knew', he recorded. Yet imagine any man, Fleay reflected, 'carrying the responsibilities Churchill did, with humanity on the rack in Europe and Asia, finding time to think about, let alone want, half-a-dozen duck-billed platypuses from faraway Australia'.

Yet no matter how 'sacred' the platypus might be, a protected species in every State, Churchill had spoken and the business would be attended to! Indeed, might not the platypuses become urgers for more planes and guns, minuscule 'beavers'—recalling the punchy Canadian, Lord Beaverbrook who, as Minister for Aircraft Production during the Battle of Britain, had successfully galvanised the women of Britain to offer up their aluminium saucepans to get more planes into the air? Whatever the motive, the platypus would go to war. And despite his amazement, David Fleay was the man for the task.

Neither Jack nor Jill were to be the answer to Churchill's call. Their day for fame came later. Nonetheless they played a timely part in introducing the masterminding officials to the animal's nervous and secretive ways. They at least succeeded in convincing the severe visitors from the Health Department that the idea of *six* captured platypuses, freshly taken from the wild and housed on a ship for transportation for Britain, was not a goer.

No living platypus had ever reached any part of Europe, though impregnated bottled specimens and skins had poured in in large consignments. Fleay, the man in the hot seat, made his conditions. He would capture several young platypuses, select a single best candidate, and design a portable home for its long sea journey. At the same time, a crew member with an affinity for animals would be given a crash course by Fleay and appointed as 'platypus keeper' for the voyage.

The experienced Fleay caught his young male platypus

from a tributary of the Yarra River on 1 April 1943, and the appropriately named, beady-eyed 'Winston' became an honoured national guest, diligently cared for and fed on a lavish bill of fare paid for by the Curtin Government.

Informed of his namesake and unaware of the difficulties involved, Churchill was impatient of delay. Visiting Washington for crucial talks with President Roosevelt in May that year, he communicated with Dr H. V. Evatt—Australia's Minister for Foreign Affairs, also on a mission in Washington —to speed things along. Evatt's terse cablegram to Australia's foreign affairs representative in Canberra for transmission to the Commonwealth Director-General of Health, Dr Cumpston, read simply: 'Churchill at Washington most anxious that platypus should leave immediately. What is present situation?'

However, it was not until September 1943, four months after his arrival at Healesville, that young Winston was ready to take up residence in the travelling platypusary—and then not without some misgivings on the organiser's part. Built with the utmost secrecy by the government, his new habitat, over and above the usual features, sported a swimming tank equipped with bafflers to overcome water roll and waterproof 'lead-in' burrows.

Late in September 1943, after seven months of covert preparation, Winston sailed from Melbourne in MV *Port Phillip* bound on his quixotic wartime assignment.

The ship was well prepared, equipped, as Fleay noted, with 'enough earthworms, crayfish, mealworms and fresh water to have refuelled Winston on a complete round the world voyage'. Merchant Vessel *Port Phillip* sailed across the Pacific, through the Panama Canal and into the submarine infested Atlantic Ocean. In Britain, Sir Winston Churchill made arrangements with the Royal Zoological Society of London for the reception of the Australian 'operative'.

Almost through the Atlantic, a thriving and healthy Winston was feeding ravenously. Within four days' sail of England, disaster struck. The ship's sonar detected the presence of a submarine. The rapid discharge of depth charges into the surrounding waters saved the ship and her men. But the jarring detonations instantly killed the platypus. His highly sensitive, nerve-pocked bill, designed as a complex sense organ to detect the smallest insect at the bottom of a river and to respond to the slightest vibrations of the natural world, was unable to deal with the violent explosions of men.

By coincidence, in October 1943, as Winston journeyed on his heroic mission, the platypus Jill began gathering material for a nesting burrow and, after mating with the now familiar Jack, started to construct her nursery under the fascinated eyes of the spectators.

After her eggs were laid, she blocked her burrow with several 'pugs' and at the end of a long wait, notable for Jill's intensive eating, Fleay and his staff dug into the burrow in January 1944 and extracted one fat wrinkled infant platypus. They named her 'Corrie'. The first platypus to be born in captivity, she was soon a major media event.

Corrie's birth resonated in Britain, and Australia gained some of the liveliest publicity since the battle of El Alamein. In the thick of war, this clearly ruffled the London *Daily Telegraph* who published a local rhymester's verse in their columns:

Hush-a-bye Platypus,
Pride of the Zoo
Baby shall figure in Nature's Who's Who,
Mummy will fondle and Daddy will brag
While all the zoologists' tongues are a-wag.

Shush, little mammal, you're not all that smart,
This is no time to expect a star part,
Sleep—and remove that smirk off your bill
We are making more history than you ever will.

'Australian jungle fighters', huffed the paper 'will doubtless agree'.

Late in 1945, at the close of World War II, Field-Marshal Lord Alanbrooke, wartime Chief of the Imperial General Staff, visited Melbourne and spent an afternoon enjoying the fauna of the Healesville Sanctuary. Alanbrooke cuddled the two-year-old Corrie. It was his first encounter with a live platypus. His only previous contact with the animal, he told Fleay, had been with a mounted specimen on Sir Winston Churchill's desk. 'Winston' had reached his destination!

15

'THE ANIMAL OF ALL TIME'

The relationship of the monotremes to the rest of the mammals continues to be a matter of debate.

Mervyn Griffiths, *The Biology of the Monotremes*, 1978

Bubbling beak,
and flapping feet.

Unsourced Aboriginal song

From the 1960s an upsurge of research interest in Australia and elsewhere added new knowledge of the physiology, ecology, biochemistry and palaeontology of the monotremes. In 1995, Tom Grant's *The Platypus* extended knowledge of platypus ecology, tracing the annual cycle of platypus life and revealing how these evolutionary survivors have adapted successfully to their modern environment in south-eastern Australia. In its annual cycle, the platypus progresses through the rigours of winter to the frenzied activities of spring mating and rearing, and the more leisurely times of summer to autumn. Grant wrote:

> For all platypuses this [autumn] season is one of preparation for winter ahead... At the end of autumn most adult platypuses are at their maximum weight, and their fur is in the best condition. For the juveniles this season is the time when many are dispersing and is probably a time preceding considerable juvenile mortality over the winter, especially when drought reduces the available habitat.
>
> The platypus as a species has been going through this annual cycle for a very, very long time, but it is only now that we are aware of what each season means to *Ornithorhynchus anatinus*.

Among the many oddities associated with the animal's physiology is surely the platypus's extraordinary capacity for consuming food. Bennett sensed this, stuffing his luckless captives with meat and worms, but he got it badly wrong. Burrell was defeated by the enormous amounts required to sustain several animals in captivity but he was the first to observe that a platypus could catch half its body weight in live prey during one night's feeding. The animal's skill in diving in the underwater darkness using only its sense of touch to locate its prey convinced Burrell that it must have a 'sixth sense' to determine its direction.

Some 70 years later (for leaps in platypus understanding were typically slow) the German physiologist Henning Scheich at the Technical University, Darmstadt was sent a pickled platypus head by the curator of the Australian National

University's zoology museum. Examining at the same time a close-up photograph of the platypus bill and noting the dense array of pores scattered on the skin surface, Scheich postulated that the pores could be sense organs similar to those found on the skin of certain marine and freshwater fish who located their prey and communicated with each other by means of electric fields. In 1985, Scheich and an assistant came to Canberra to explore his hunch with colleagues at the Australian National University.

German interest in the platypus had flickered strongly across the years. Early in the nineteenth century, as debate on the platypus quickened, Meckel and Baer had pushed out pioneering anatomical frontiers. Seventy years on, zoologist and embryologist Richard Semon, trained by Ernst Haeckel in Jena, spent time during 1892–93 on Queensland's Burnett River close to the site of Caldwell's investigation, where—like the Scottish biologist before him—he turned his gaze and disciplined energy to the study of the lungfish, *Ceratodus* and 'oviparous mammals'.

Like Caldwell, Semon relied heavily on the help of the Aborigines, bagged a horde of the *Ornithorhynchus*, made caps from their fur 'as a nice keepsake from the antipodes' for friends at home, and recorded observations on the platypus's teeth, the range of body temperature in monotremes, the habits of echidna and platypus and their long survival across geological time, in his book *In the Australian Bush*.

A century after Semon, Scheich in Australia stuck to the laboratory. Placing a platypus in a large tank, he found that the animal would dive to the bottom and overturn bricks that concealed live batteries while ignoring bricks hiding batteries that were flat. In a second experiment, the platypus was confronted with a large perspex screen suspended in the water which shielded a live battery. Swimming in its customary way with eyes, ears and nasal cavity closed, the animal dived eagerly around the screen—but bumped into it unsuspectingly when the battery was flat.

The 'platypus electric'—a monotreme furnished with unique electro-reception—was unveiled!

Scheich's flashpoint of discovery revealed that the platypus used its 'electric sense' under water both for avoiding obstacles and for finding food in its murky habitat. So sophisticated was this 'sixth sense', the animal could detect prey tucked away in mud and under rocks. It was a far cry from the general belief that the platypus simply blundered along the bottom of the stream seizing whatever food it could.

All animals generate some electricity in their contracting muscles when they make movements. Some, like fish and amphibians, also use an electrosensory system for generating active pulses for communication or in the case of the electric eel for producing an electric field strong enough to stun its prey. By contrast, these findings indicated that the platypus was a 'passive electrodectector', sussing out its prey through electric sensors on its bill. While sight and hearing played no part in its vigorous filling of its cheek pouches, the platypus found its diverse cuisine at a distance by using 'electrolocation'. Its favourite food, the freshwater shrimp, popped into the experimental tank, was found to generate an electric field through its tail flip large enough to be detected by a platypus 10 centimetres away.

Most significantly for science, the experiment indicated

that electrical stimulation of the platypus bill worked—quite differently from that of fish or the sonar wave electrolocation used by whales and bats—by evoking electrical responses from the surface of the platypus brain.

The bold announcement of the discovery with its cover page illustration 'Electrolocation by Duck-Billed Platypus' in the international weekly journal of science, *Nature*, was greeted with amazement and some disbelief. That quirky Australian animal had done it again! Here was another shock for science. The discovery, declared one Australian curator of mammals firmly, ranked as 'one of the major natural history discoveries of this era in the world'.

Stirred by the findings, researchers in Victoria swung into action. At Monash University in Melbourne a group of colleagues led by physiologist Uwe Proske took up the challenge. Many moonlit nights on the banks of local streams later, they had the elusive animal in their net. Examining it under anaesthetic, they soon discovered from nerve fibre recordings that the mucous sensory glands under the duckbill's skin were the site from which electrical activity flowed.

These electroreceptors were not, it proved, randomly distributed across the bill but lay in lines over the shield. Ever since the animal's first dramatic appearance in London nearly two centuries earlier, this curious flap of skin extending from the base of the duckbill's upper and lower jaws had puzzled and vexed observers. Its very presence suggested that the animal was a hoax. Mother Nature would never make so botched and untidy a join!

The role of the offending shield, says Proske, long remained a subject of speculation. Yet, with the many receptors he and his research team discovered on it, he speculated that the shield probably helped provide the platypus 'with a three-dimensional view of its electric world'.

Research on the electric monotreme opened unexpected new doors. In certain marine and freshwater fish, their electric receptors are known to consist of a specialised hair cell that communicates with the skin surface. Here the receptor cell responds to a stimulating voltage by secreting a chemical which excites the nerve fibres. In the case of the platypus, the electric stimulus excites the nerve fibre directly and there is no chemical mediator.

Excited researchers now grasped the fact that not only had nature produced at least two distinctly different electrosensory systems in the evolutionary scheme but, in Proske's words, 'if an entire sensory system has evolved *de novo* in the platypus, it must be considered a highly evolved animal and not just a primitive transition between reptiles and mammals'. 'We believe', Proske wrote elsewhere, 'it left the mainstream of mammalian evolution a long time ago, long enough to have evolved a completely new sensory system'.

Thus, far from the old notion of Australia as a mammalian penal colony—a faunal Gulag—here, as Australia's leading monotreme authority, Mervyn Griffiths, expressed it, was 'the animal of all time'.

At Sydney's Taronga Park Zoo, an exuberant curator of mammals, Dr Deedie Woodside, spelled out the message. 'Australians', she told the local press in November 1985, 'have long regarded their animals as a group which is dying out, almost something we have to apologise for. What we have here is evolution in isolation—a mammal developing this trait in Australia, independently of anything else. They're not dying out, they're becoming more and more specialised. It is very exciting.'

The platypus had been tagged by Darwin as a key exemplar of the principle of evolutionary adaptation in isolation, and now, over a century after *The Origin of Species*, his point had been dramatically established.

Yet over what vast periods of geological time had this process of adaptation occurred? What do we know about the evolution of this specialised and resilient monotreme?

Australia, contained within the southern landmass Gondwana, which broke away from the earth's continents some 225 million years ago, was first joined with Africa, South America, Antarctica, India and Madagascar. Linked, and later alone, she drifted through a long series of changing climatic zones that shaped the evolution of her unqiue fauna and flora.

The piecing together of this evolution has been slow and arduous. The fossil record has only recently offered up its fragmentary but telling clues.

In 1987 the vivid dry orange country of Lightning Ridge in the opal region of Walgett, New South Wales, yielded a small fragment of opalised jaw containing three highly distinctive teeth that revealed it to be from a monotreme. Gleaming in the brilliant sunlight, 'the world's only transparent fossil' declared the press, it was the relic of Australia's oldest fossil mammal, a remote large-sized platypus ancestor from the Cretaceous period some 100 million years ago in the Mesozoic era of geological time.

This animal would have inhabited a landscape of broad river estuary and inland sea very different from Lightning

Ridge's low arid hills today. It would have swum at the feet of herbivore dinosaurs. Long after the bones of this distant forebear were encased in sedimentary rock, the land was uplifted and a silica-rich groundwater, seeping through the sediment, dispersed the bones and replaced them perfectly with opal. The extinct creature was named *Steropodon galmani*, its fiery teeth suggesting a monotreme precursor of great antiquity, significantly larger and stumpier than the present platypus but sharing many traits including a flesh-covered bill, a distinctive posture, very small eyes and the absence of an external ear.

A decade earlier, in 1975, two fossil teeth of a more recent platypus predecessor from the Miocene epoch, some 15–25 million years ago, were found in deposits of an ancient lake bed in South Australia. Morphologically akin in root and form to those of the living juvenile platypuses, the fossil teeth were larger and less complex. In the present-day platypus, teeth are only found in the very young animal and are replaced in adulthood by the large horny pads which serve as efficient grinders for the food. The extinct creature was named by palaeontologists *Obdurodon insignis*, 'significantly enduring teeth'.

Subsequently an unearthed leg bone fragment from the first close relative of the modern platypus was dated to some 4.5 million years ago. And, as this sentence issues from my wordprocessor in early April 2000, the Australian media has announced the finding of a small leg bone, similar in size and shape to that of the living platypus, dredged from the sediments of Queensland's Darling Downs and evidencing an age of between 20 000 and 200 000 years.

Steropodon galmani, *a Cretaceous platypus from Lightning Ridge, New South Wales. A palaeontological and artistic reconstruction of an extinct forebear by Peter Schouten. Such a reconstruction is not intended to be the last word on how the extinct animal appeared. It is offered as a hypothesis reflecting current understanding and could be updated as new findings come to light.*

One thing seems certain. The monotremes, as Griffiths writes, are 'a taxon of mammals', of Mesozoic origin which has survived the vicissitudes of millions of years of existence during which their relatives have gone to the wall'. Genetic data suggests that the ancestors of the platypus 'split' from the line which evolved into its monotreme relative, the echidna, some 60 million years ago.

For at least 40 000 years—some anthropologists calculate up to 60 000 or more—the Aborigines have inhabited Australia. Their knowledge of the platypus has been both practical and totemic. Over long centuries, they have known its singular habits and physiognomy, been both delighted or repelled by it as a source of food, and have woven it into their Dreaming stories.

During 1844, an enlightened Protector of Aborigines in the Colony of Victoria, George Angus Robinson, travelling about his domain and documenting the language and culture of the Indigenous people, recorded a Creation story that linked the platypus with the formation of the great Snowy River which joins the Monaro country of New South Wales with Victoria. It ran:

> The Moon made the rivers, took a large quantity of sea water to the mountains beyond Maneroo, i.e. Snowy Mountains. On its journey among the mountains it was scented by the Water Mole [platypus] which smelt the water when the Moon rested. The Moon went a long way and the Water Mole still tracked on and finding the Moon asleep

struck a yam stick into the water, where it gushed out and formed the river, and the Moon was thus 'kubbah big sulky'.

At the far northern tip of the continent, in the dry regions of Cape York, the recently revealed prehistoric rock art of the Quinkan shows a platypus figure painted in linear style in dark red dry pigment. The drawing is in a well-protected site along with a diprotodont and a koala—all three animals are now extinct in the region. The platypus's placement in the artwork suggests that this was a haunt of the animal in an earlier period and that the platypus had a place in local Aboriginal Dreaming.

Teachings from the Dreaming, for Aboriginal people, explain physical and spiritual phenomenon and the rules governing interrelationships between people, land and spiritual beliefs. They are linked with totemism by which the clan or language group have selected living creatures such as the kangaroo, the emu, the goanna, the lizard, the snake and the fish, as specially named companions and totem guides. In practice, a clan must select a totem before it can occupy a hunting ground and these totems are woven as key presences into Aboriginal Dreamings.

In the relationships of totems to one another, the unique character of the platypus placed it in an unusual position in the family of animals. The following tale from the Berrwerina [Brewarrina] people of the Darling River in New South Wales recorded by the famous Aboriginal teacher and author, David Unaipon, reveals the long thread of Aboriginal knowledge of the platypus and its ways.

There was a time, Unaipon's story goes, 'when the Animals, Birds and Reptiles multiplied so greatly in numbers that the country in which they lived was not large enough to accommodate them'. The Kangaroo and 'Teddy Bear' [koala] assembled with the Bird tribes and the Goanna and Tiger Snake who represented the reptiles to confer on this difficult overcrowding. The Kangaroo introduced the Platypus to the assembly. 'He is present here this day', he declared, 'because he belongs to a class that has a majority in numbers. His family is more than the whole of my race and your family, oh bird tribe, and placing your family with ours, oh reptile tribe, this gentleman outnumbers us still'. The Kangaroo asked the reptiles and birds, could they ask the platypus to take some means to prevent their rapid increase. 'Just look on either side of this range of mountains—platypuses in great numbers are to be seen everywhere!'

Impatient at the long conference, the Frilled Lizard whose totems were the elements—Lightning, Thunder, Rain, Hail & Wind—called upon them to bring on a great deluge to destroy the platypus family. They had become too numerous and were more easily overtaken in flood than other families.

Great dark clouds mantled the clear sky, the lightning flashed, the thunder roared and the winds came howling, driving the rain and hail into every hiding place of animal, bird and reptile. The animals were able to hide from the storm but the Platypus were all killed. The Platypus were no more.

Three years later, the Carpet Snake, acting on a report from the Cormorant, reported to the animals that he had found Platypus hiding in a quiet distant mountain stream. Kangaroo called a great meeting.

The question now arose [Unaipon tells] as to who Platypus's relatives were so that they would be more likely to help them in future. Black duck could see a resemblance to widgeon duck, which has comb-like teeth along the edge of its bill, so the birds swore to help the platypus. The pelican added that he must be related because he laid eggs. The reptiles could see no real relationship, except possibly the egg. The animals thought they were related, but could not see how.

Kangaroo offered Platypus the most beautiful bride from animal, bird or reptile. A long discussion followed as to the correct relationship of Platypus to the others. But Platypus chose the Bandicoot as the tribe to which he should be allied. He chose a young Bandicoot wife and 'hence was related to the hairy tribe'. But one of these, the Water Rat, became jealous, and there was a great battle which the Platypus won with the assistance of the Bandicoot wives.

So [runs the story] the Platypus, although he has tried to break away from the bird tribe, is ever reminded that his wife lays eggs and still retains the bird or duck's bill. He has tried hard to separate himself and family from the birds. He has succeeded in getting rid of the feathers that the Emu reminded him of, that he wore long and long ago, and he tried to sever his connection with the animals. But in this he has also failed. He still belongs to the Kangaroo Possum tribe. He is making a desperate effort to cause a great gulf between himself and the kindly Kangaroo, who looks upon him with pity. The more he tries the greater becomes the difficulty. So he makes no more effort, but simply contents himself to be neither bird, reptile, nor kangaroo, but a plain platypus.

He no longer seeks the companionship of the animal, bird, or reptile families, but lives affectionately with his wife in the rivers and billabongs. In songs he petitions the God of Food:

Give me more grubs as the water ripple
And make for me bread of the Nardoo seed
Do this often and often as the sun do rise;
And I shall gather in the shade of night
The Food thou shalt throw into the water hunting place
    of mine
Because I'm hiding all the day fearing that someone may
    call up the Lightning Thunder and Rain.

Unaipon's Dreaming story—an oral legend of the Aboriginal people—was to find an unexpected outlet some 30 years before it saw print, in an early Australian book for children, *Dot and the Kangaroo*, published in 1899. Its author, Ethel Pedley, a talented musician, often spent time visiting her brother at his pastoral property near Walgett, New South Wales, where she grew to know the Australian bush and its creatures and to gain knowledge of Aboriginal people. Combining this with a clear grasp of evolutionary principles and a nod to Christianity, she created the story of the small heroine Dot, lost in the tangled green forest, which was to become a classic of Australian literature.

A female kangaroo, searching for her own lost baby joey, stops to help Dot find her way home, but not before giving her some special 'berries of understanding' to eat, which allowed her to communicate with the animals.

'I've never seen a Platypus,' said Dot. 'Do tell me what it is like!'

'I couldn't describe it,' said the Kangaroo, with a shudder. 'It seems made up of parts of two or three different sorts of creatures. None of us can account for it. It must have been an experiment, when all the rest of us were made; or else it was made up of the odds and ends of the birds and beasts that were left over after we were all finished.'

Little Dot clapped her hands. 'Oh, dear Kangaroo,' she said, 'do take me to see the Platypus! There was nothing like that in my Noah's Ark.'

'I should say not!' remarked the Kangaroo. 'The animals in the Ark said they were each to be of its kind, and every sort of bird and beast refused to admit the Platypus, because it was of so many kinds; and at last Noah turned it out to swim for itself, because there was such a row. That's why the Platypus is so secluded. Ever since then no Platypus is friendly with any other creature, and no animal or bird is more than just polite to it. They couldn't be, you see, because of that trouble in the Ark.'

'But that was so long ago,' said Dot . . . 'and, after all, this is not the same Platypus, nor are all the bush creatures the same now as then.'

'No,' returned the Kangaroo, 'and some say there was no Ark, and no fuss over the matter, but that, of course, doesn't make any difference, for it's a very ancient quarrel, so it must be kept up. But if we are to go to the Platypus

The Platypus sings to Dot and the Kangaroo of its antediluvian days.

we had better start now; it is a good time to see it—so come along, little Dot,' said the Kangaroo.

Finding the Platypus beside the stream, Dot and the Kangaroo listen as it sings to them of its antediluvian forebears.

'It breaks my heart to think that they are all fossils,' it exclaimed, mournfully shaking its head. 'Fossils!' it repeated, as it plunged into the pool and swam away. 'Fossils!'

The platypus continues to lead its quiet life in the rivers of eastern Australia. Rippling its way in a thousand streams on far mountain slopes, in creeks and reservoirs, near towns and cities, it remains private and elusive. It inhabits the inland river systems and their tributaries that stretch from Cooktown in North Queensland through the Great Dividing Range and major rivers of New South Wales, to the Murray–Darling basin into Victoria and on to Tasmania; adapted both to tropic streams and rivers; to the cold alpine waters of the Snowy Mountains and, increasingly, to urban river sites near Melbourne. A small handful of platypuses were taken from the mainland to Kangaroo Island between 1928 and 1946 and their descendants are the sole remaining group to be found breeding in the wild in South Australia.

Although ecologically vulnerable to pollution in waterways from garbage, fishing lines and nets, and agricultural development, the platypus is not rare—although, with its covert twilight ways, sightings in the wild can be difficult. An air-breathing mammal, it spends some seventeen hours a day out of the water resting in its underground burrow and is rarely observed on land except for a few minutes. The male, an intrepid adventurer in search of food—and sex in the mating season—can travel up to eleven kilometres in a night with a spatial memory that allows him to navigate over long distances. He lives for approximately five years. The female keeps to a

The platypus uses its two-holed, touch-sensitive bill as its underwater
'eyes and ears' (which are closed during its underwater sorties) to search
the river bottom for food.

smaller environmental range and lives to over eight years and sometimes considerably more. Both male and female are essentially solitary.

Research continues to open up knowledge of the physiology and habits of this small engaging creature, now internationally recognised as an Australian icon. The questions that puzzled earlier researchers on what happened in the nesting burrow have been deciphered. The female incubates her one to two (and sometimes, but rarely, three) eggs in her nesting burrow for ten or eleven days and suckles the young until they leave the burrow three to four months after birth, well-furred and able to swim. The fine fur, made up of some 900 hairs covering each square millimetre of skin, has two layers—a woolly undercoat and shiny longer guard hair—which together trap a layer of air next to the platypus skin, keeping most of its body dry when diving.

The contentious spur located on each hind leg of the male is now conclusively known to secrete venom which is used by mature males in the breeding period when competing for mates and territories. It is sometimes used as a defensive weapon against humans or dogs when shot or caught and research goes on in search of an antivenene. The platypus is thus the only Australian mammal known to be venomous.

The platypus is now protected by law throughout Australia and is not allowed to be taken overseas for any purpose. Conservation guidelines, urban platypus programs, the tagging, monitoring, radio-tracking and reporting of platypus sightings, and research and education programs involving participants of all ages and designed to ensure the preservation of this unique mammal are conducted by the Australian Platypus

Conservancy in Victoria, Earthwatch Australia and community groups in Tasmania and Queensland.

While information and understanding of the platypus expands, more research remains to be done. Yet one point emerges clearly: the platypus, with its strange melange of characters and affinities associated with mammal, reptile and bird, is *not*—as nineteenth-century theorists and many twentieth-century biologists insisted—a 'primitive' animal. Instead, as biologist Stephen Jay Gould asserts, it is an animal 'actively evolving in its own interests'.

A new century for the platypus has arrived. Wonder remains, and delight. Australians have adopted the little animal as their talisman. Swimming and diving quietly across time, the platypus smiles.

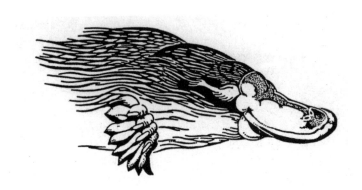

GLOSSARY

Bruta: the lowest Linnean order which included what we now know as the Edentata and to which Shaw originally consigned the platypus.

classification: the process of ordering organisms into groups on the basis of their relationship; and the product of such ordering. The hierarchy of categories in the classificatory system is: phylum, class, order, family, genus (plural genera) and species. The platypus is classified as class Mammalia; subclass Prototheria; order Monotremata; family Ornithorhynchidae; genus and species *Ornithorhynchus anatinus*.

Edentata: a Mammalian order of toothless animals comprising living and extinct anteaters, sloths and armadillos, to which—it was once considered—that the platypus might belong.

Eutheria: a subclass of mammals encompassing the dominant vertebrates, live-bearing (viviparous) mammals which nourish a foetus by means of a placenta and give birth to young which are fully formed. Eutherian mammals include humans, and, for example, horses, deer, elephants, koalas and cats. They are more commonly referred to as placental mammals.

mammals (class Mammalia): members of a class of vertebrate animals, the females having mammary glands (mammae) producing milk from which they nurse their young. The three subclasses of mammals are Prototheria, Metatheria or Marsupialia, and Eutheria.

marsupials: subclass Marsupialia or Metatheria. Their most striking feature is that they are born in an embryonic condition and are brought to development by nourishment on the mother's teat in the pouch.

monotremes: subclass Prototheria. An order of mammals who have separate uteri entering a common urino-genital passage joined to a cloaca into which the gut and excretory systems also enter. Like all other mammals, monotremes have seven cervical vertebrae, but unlike those of Eutherians and Metatherians (marsupials), they bear cervical ribs.

oviparous: egg-laying—that is, producing young by means of eggs expelled from the body before being hatched.

ovoviviparous: producing fully formed eggs that are retained and hatched inside the mother's body before the young are expelled.

phylum (plural phyla): a taxonomic division comprising a number of classes.

Prototheria: the name given by Lamarck to a new class which included the platypus and echidna, and which remains a subclass of the Mammalian class today.

taxon: the scientific name of a category or group of organisms of any rank sufficiently distinct from other groups of organisms to be worthy of being assigned a separate category.

taxonomy: the theory and practice of classifying organisms according to their resemblances and differences. Taxonomy functions as an information storage and retrieval system in zoology and botany.

vertebrates: animals possessing a bony skeleton and a well-developed brain. They are now divided into five classes—fishes, amphibia, reptiles, birds and mammals. At the time of the discovery of the platypus, there were four classes only, amphibians then being included with reptiles.

viviparous: bringing forth live offspring from within the body of the parent.

Measurement conversions

1 inch	25.4 millimetres
12 inches (1 foot)	30.5 centimetres
3 feet (1 yard)	0.914 metres
1 mile	1.61 kilometres

A WORD ON
SOURCES

The story of the platypus and its long and intriguing presence in the history of science first fascinated me when I wrote *Scientists in Nineteenth Century Australia: A Documentary History* (Cassell Australia, Sydney, 1976) and, again, when I drew together a broad and illustrated history of Australia's nineteenth-century science in *'A Bright & Savage Land': Scientists in Colonial Australia* (Macmillan, Sydney, 1986). So, when in 1998, Allen & Unwin's science publisher, Ian Bowring, invited me to consider a narrative topic for the general reader involving the history of Australian and international science, I chose the 'platypus mystery'.

Well embarked on my research and writing, I came upon two scholarly works—deposited in somewhat inaccessible places—both written by American authors, which attended to aspects of the platypus story from a specifically academic perspective.

The first was Kathleen G. Dugan's 'Marsupials and Monotremes in Pre-Darwinian Theory', a PhD thesis from the Department of History, University of Kansas, 1980; the second was Jacob W. Gruber's 'Does the Platypus Lay Eggs? The history of an event in science', an erudite paper published in *Archives of Natural History*, vol. 18, no. 1, 1991, pp. 51–123.

Both works have proved a considerable pleasure and assistance to me and indicated new sources of information, and I extend my thanks to the authors. I am in particular debt to Jacob Gruber for information quoted or drawn from his paper, pp. 55–7, 64–5, 72–9, 80–1, 84, 104 and bibliography. A further recent work, Harriet Ritvo, *The Platypus and the Mermaid* (Harvard University Press, Cambridge, Massachusetts, 1997), encompasses a diverse and stimulating canvas and I have drawn material from pp. 5, 7, 15, 18 and 50.

Harry Burrell, *The Platypus*, (Angus & Robertson, Sydney, 1927) is a foundation work which has provided both quoted and other source material, as has George Bennett, *Gathering of a Naturalist in Australasia: Being Observations Principally on the Animal and Vegetable Productions of New South Wales, New Zealand, and Some Austral Islands* (John van Voorst, London, 1860), and my own books mentioned above.

General Sources
The following books and articles furnished broad reading and, where pages are noted, quotations and specific information. The relevant scientific journal articles of George Bennett, Richard Owen, Geoffroy St-Hilaire and J. T. Wilson have been consulted but are not listed.

Appel, Tony A., 'Henri de Blainville and the Animal Series: A Nineteenth Century Chain of Being', *Journal of the History of Biology*, vol. 13, no. 2, Fall 1980, pp. 291–319

Bennett, George, *Wanderings in New South Wales, Batavia, Pedir Coast, Singapore and China: Being the Journal of a Naturalist in those Countries During 1832, 1833 and 1834*, 2 volumes, London, 1834

Bewick, Thomas, *A General History of Quadrupeds*, 4th edn., Newcastle upon Tyne, 1800, pp. 521, 525

Buffon, Baron (Georges), *Natural History*, abridged by the Rev. W. Hutton, vol. 1, J. Tegg, London, 1821

Burkhardt, Frederick & Sydney Smith, *The Correspondence of Charles Darwin*, Cambridge University Press, Cambridge, 1987–1993: vols 3, p. 25; 5, pp. 232, 234, 247–9; 7, p. 345; 8, p. 349

Clarke, F., *The Land of Contrarieties*, Melbourne University Press, Melbourne, 1997, p. 156

Coppleson, V. M., 'The Life and Times of Dr. George Bennett: Annual Post Graduate Orations', *Bulletin of the Post-Graduate Committee in Medicine*, University of Sydney, ii, 1955, pp. 251–61

Cunningham, Peter, *Two Years in New South Wales, (1827)*, edited by David S. Macmillan, Angus & Robertson, Sydney, 1966, p. 160

Cuvier, Baron (Georges), *The Animal Kingdom*, translated by Edward Griffith, Charles Hamilton Smith and Edward Pidgeon, George Whittaker, London, 1817

Darwin, Charles, *On the Origin of Species*, John Murray, London, 2nd edn, 1861, pp. 113, 344, 368

——*Descent of Man*, John Murray, London, 1874, pp. 243–4, 255

Darwin, Francis (ed.) *Life and Letters of Charles Darwin*, John Murray, London, 1888, vol. 2, pp. 335, 339–40

de Beer, Gavin, *Charles Darwin: A Scientific Biography*, Double Day, Anchor Books, New York, 1965

Desmond, Adrian, *Archetypes and Ancestors: Palaeontology in Victorian London, 1850–1875*, University of Chicago Press, Chicago, 1982, pp. 42, 47, 453

——and James Moore, *Darwin*, Michael Joseph, London, 1991

Dugan, Kathleen G., 'The Zoological Exploration of the Australian Region and its Impact on Biological Theory', in *Scientific Colonialism: A Cross-Cultural Comparison*, edited by Nathan Reingold and Marc Rosenberg, Smithsonian Press, Washington, 1997, pp. 79–100

Eco, Umberto, *Kant and the Platypus: Essays on Language and Cognition*, translated by Alistair McEwen, Secker & Warburg, London, 1999, p. 245

Field, Barron, *First Fruits of Australian Poetry*, Sydney, 1819, 2nd edn, 1823

——(ed.), *Geographical Memoirs of New South Wales: by various Hands*, London, 1825

PLATYPUS

Fleay, David, *We Breed the Platypus*, Robertson & Mullens, Melbourne, 1944

——*Paradoxical Platypus*, Jacaranda Press, Brisbane, 1980, Chapters 6 and 7

Gould, Stephen Jay, *Bully for Bronotsaurus*, W. W. Norton & Co., New York, 1991, pp. 272–8

Grant, Tom, *The Platypus: A Unique Mammal*, University of New South Wales Press, Sydney, 1995

Griffiths, Mervyn, *The Biology of the Monotremes*, Academic Press, London, 1978

Hoare, Michael E., 'All Things are "Queer and Opposite": Scientific Societies in Tasmania in the 1840's', *Isis*, vol. 60, 1969, pp. 363–6

Horner, Frank, *The French Reconnaissance: Baudin in Australian Waters, 1801–1803*, Melbourne University Press, Melbourne, 1987

Hunt, Susan & Paul Carter, *Terre Napoleon: Australia through French Eyes, 1800–1804*, Catalogue of an Exhibition in collaboration with the Collection Charles-Alexandre Lesueur of Museum d'Histoire Naturelle, from Ville du Havre, France, held at the Museum of Sydney, 27 February–30 May 1999

Huxley, T. H. 'Essay on Owen's Position in the History of Anatomical Science', in Rev. Richard Owen, *The Life of Richard Owen*, John Murray, London, 1894, vol. 2, pp. 271–332

Keynes, Richard Darwin (ed.), *Charles Darwin's Diary of the Voyage of H.M.S. Beagle*, Cambridge University Press, Cambridge, 1933

King, P. P., *Narrative of a Survey of Intertropical Waters and Western Coasts of Australia Performed Between the Years 1818 and 1822*, John Murray, London, 1827.

Macleod, Roy M., 'Evolution and Richard Owen, 1830–1868: An Episode in Darwin's Century', *Isis*, vol. 56, 1965, pp. 259–80

Morison, Patricia, *J. T. Wilson and the Fraternity of Duckmaloi*, Editions Rodopi, Amsterdam, The Welcome Institute Series in the History of Medicine, 1997, Chapter 7 and Bibliography

212

Mozley Moyal, Ann, 'Evolution and the Climate of Opinion in Australia, 1830–1876', *Victorian Studies*, University of Indiana, vol. 10, no. 4, 1967, p. 411–30

——'Sir Richard Owen and his Influence on Australian Zoological Science', *Records of the Australian Academy of Science*, vol. 3, no. 2, 1975, pp. 41–55

Newland, Elizabeth D., 'Dr. George Bennett and Sir Richard Owen: A Case Study of Colonization of Early Australian Science', in R. W. Home and S. G. Kohlstedt (eds), *International Science and National Scientific Identity: Australia Between Britain and America*, Kluwer Academic Publications, Boston, 1991, pp. 55–74

Nicholas, F. W. & J. M., *Charles Darwin in Australia,* Cambridge University Press, Cambridge, 1989

Owen, Richard, 'Remarks on the "Observations sur L'Ornithorhynque" par M. Jules Verreaux', *Annals and Magazine of Natural History*, vol. II, Series 5–2, 1848, pp. 317–22

——Correspondence from his Papers, Australian Joint Copying Project, National Library of Australia, and Mitchell Library, Sydney

Pedley, Ethel C., *Dot and the Kangaroo*, illustrated by Frank P. Mahoney, T. Burleigh, London, 1899, pp. 21–2; many later editions

Péron, M. F., *A Voyage of Discovery to the Southern Hemisphere Performed. . . During the Years 1801, 1802, 1803, 1804*, translated from the French, London, 1809

Proske, Uwe, 'The Monotreme Electric', *Australian Natural History*, vol. 3, no. 4, 1990, pp. 289–95

——'The Platypus: An Electric Hunter', *Today's Life Science*, August 1994, pp. 44–50

Ripples, Newsletter of the Australian Platypus Conservancy, Whittlesea, Victoria, 1998–2000

Scheich, H., G. Langer, C. Tidemann, R. B. Coles & A. Guppy, 'Electroreception and Electrolocation in Platypus', *Nature*, vol. 319, 1986, pp. 401–2

Semon, Richard, *In the Australian Bush*, Macmillan, London, 1899; English edition

Smith, Bernard, *European Vision and the South Pacific 1768–1859*, 2nd edition, Harper & Row, London, 1985

Strahan, Ronald (ed.), *The Australian Museum Complete Book of Australian Mammals*, Angus & Robertson, Sydney, 1983

Strzelecki, Paul E. de, *Physical Description of New South Wales and Van Diemen's Land*, Longman, Brown, Green & Longmans, London, 1845, pp. 259–60

Unaipon, David, Unpublished manuscript of his legends, Mitchell Library, Sydney. These were published without acknowledgment by William Ramsay Smith as *Myths and Legends of the Australian Aboriginals*, George Harrod, London, 1930

Webb, Joan, *George Caley: Nineteenth Century Naturalist: A Biography*, Surrey Beattley, NSW, 1995

Whitley, Gilbert, *More Early History of Australian Zoology*, Zoological Society of New South Wales, Sydney, 1975

Williams, G. A. & M. Serena, *Living with Platypus*, Australian Platypus Conservancy, Whittlesea, Victoria, 1999

ILLUSTRATIONS

I acknowledge, with thanks, permissions received from the following libraries, institutions, publishers and individuals to publish the illustrations contained within this book.

27 Engraving by Choubard after a drawing by C. A. Lesueur, from *Baudin in Australian Waters*, op. cit. National Library of Australia

36 Baron (Georges) Buffon, *Natural History*, vol. 1, J. Tegg, London, 1821, Plate 22, p. 209. National Library of Australia

64 Illustration by Ferdinand Bauer, in Robert Brown, *Prodromus Florae Novae Hollandiae et Insulae Van Diemen...*, London, 1813. National Library of Australia

65 Watercolour by John Lewin, 1802. Mitchell Library, ML896, State Library of New South Wales

83 Sketch by George Bennett in 'Notes on the Natural History of the *Ornithorhynchus paradoxus*, Blum', *Transactions of the Zoological Society of London*, vol. 1, 1835, p. 258. National Library of Australia

85 Woodcut from a sketch by George Bennett, in his *Gatherings of a Naturalist in Australia...*, Jan van Voorst, London, 1860, p. 118. National Library of Australia

91 From Michael Allin, *Zarafa*, Headline Book Publishing, London, 1998, p. 127

93 Engraving from a drawing by Rymer Jones, in Richard Owen, 'On the Young of the *Ornithorhynchus paradoxus*, Blum', *Transactions of the Zoological Society of London*, vol. 1, 1835, p. 228. National Library of Australia

100 From Rev. Richard Owen, *The Life of Professor Owen*, vol. 1, John Murray, London, 1894, p. 319. National Library of Australia

103 Watercolour by George Richmond, Down House, Sussex, 1840, from Alan Moorehead, *Darwin and the Beagle*, Penguin Books, Hammonsworth, Middlesex, 1969. National Library of Australia

118 From *Records of the Australian Museum*, vol. xi, 1916. National Library of Australia

120 Oil painting by Henry Williams, c. 1839. Tasmanian Museum and Art Gallery, Hobart

Colour plates

Plate

5   Oil painting by T. Richard Browne. From Mitchell Library, Sydney, and *The Skottowe Manuscript: Thomas Skottowe's Select Specimens from Nature...Newcastle, NSW, 1813*, David Ell Press, Hordern House, Sydney, 1988

6   Oil painting by John Glover of 'Ben Lomond from Mr Talbot's property', Tasmania. From John McPhee, *The Art of John Glover*

7   Illustration by John Gould and H. C. Richter, in John Gould, *The Mammals of Australia*, the author, London, 1863. National Library of Australia

8   Dave Watts / Nature Focus

INDEX